Advanced Mathematics

LAPLACE TRANSFORMS

- Theory of transformation and inverse transformation

- Transformation of elementary and complex functions

- Equations of motion of material bodies, deflection, stress, and strain of elastic beams

- Electrical current flow in network circuitries of active elements;

- Applications on heat flow equations in various geometries

- Solving partial differential equations by the operational algebraic transformations

By

Mohamed F. El-Hewie

TABLE OF CONTENTS

CHAPTER 3
ELECTRICAL APPLICATIONS OF THE LAPLACE TRANSFORMATION

CHAPTER 4
DYNAMICAL APPLICATIONS OF LAPLACE TRANSFORMS

CHAPTER 5
STRUCTURAL APPLICATIONS

CHAPTER 6
USING LAPLACE TRANSFORMATION IN SOLVING LINEAR PARTIAL DIFFERENTIAL EQUATIONS

INDEX

INTRODUCTION

This is a revised edition of the chapter on **Laplace Transforms**, which was published few years ago in Part II of My Personal Study Notes in advanced mathematics. In this edition, I typed the cursive scripts of the personal notes, edited the typographic errors, but most of all reproduced all the calculations and graphics in a modern style of representation.

The book is organized into six chapters equally distributed to address:

(1)   The theory of **Laplace transformations** and **inverse transformations** of elementary functions, supported by solved examples and exercises with given answers;

(2)   Transformation of more complex functions from **elementary transformation**;

(3)   Practical applications of Laplace transformation to **equations of motion** of material bodies and deflection, stress, and strain of elastic beams;

(4)   Solving equations of temporal **electrical current flow** in network circuitries;

(5)   Solving heat flow **equations** through various geometrical bodies; and

(6)   Solving **partial differential equations** by the operational algebraic properties of transforming and inverse transforming of partial differential equations.

During the editing process, I added plenty of comments of the underlying meaning of the arcane equations such that the reader could discern the practical weight of each mathematical formula. In a way, I attempted to convey a personal sense and feeling on the significance and philosophy of devising a mathematical equation that transcends into real-life emulation.

The reader will find this edition dense with graphic illustrations that should spare the reader the trouble of searching other references in order to infer any missing steps. In my view, detailed graphic illustrations could soothe the harshness of arcane mathematical jargon, as well as expose the merits of the assumption contemplated in the formulation.

In lieu of offering a dense textbook on Laplace Transforms, I opted to stick to my personal notes that give the memorable zest of a subject that could easily remembered when not frequently used

Mohamed F. El-Hewie

April 15, 2013

CHAPTER 1

THE LAPLACE TRANSFORMATION AND INVERSE TRANSFORMATION

1.1. Integral Transforms

The integral transform $\bar{f}(p)$ of a given function $f(x)$ in the range (a , b) is defined as follows.

$$\bar{f}(p) = \int_a^b f(x).k\,(p,x)\,dx \qquad (1)$$

Where, $k(p, x)$ is the **kernel** of the transform and is a known function of p and x. This is on the assumption that the integral exists.

The integral transform in equation (1) signifies the existence of **partial differential equations** with independent variables p and x, which upon integration over the range $x = a$ to $x = b$, yield the solution $\bar{f}(p)$.

Such differential equations possess the kernel $k(p, x)$ which take many forms when integrated over the range $x = 0$ to $x = \infty$, as follows.

Table 1. Integral Transforms of $f(x)$ to $\bar{f}(p)$ via integrating over the range $x = 0$ to x $= \infty$, after multiplication with one of five kernels given below

	k(p, x)	Name of Transform	Nature of transform
(i)	e^{-px}	Laplace transform	Exponential decay kernel
(ii)	$\sin(px)$	Fourier sine transform	Sinusoidal oscillating kernel
(iii)	$\cos(px)$	Fourier cosine transform	Sinusoidal oscillating kernel
(iv)	$x\,J_n\,(px)$	Hankel transform	Bessel function of first kind order n. Decaying oscillating kernel
(v)	x^{p-1}	Mellin transform	

Table 1 is reproduced in more details as follows

$$\bar{f}(p) = \int_0^\infty f(x).e^{-px}\,dx \qquad (2) \qquad \text{Laplace Transform}$$

$$\bar{f}(p) = \int_0^\infty f(x).\sin(p\,x)\,dx \qquad (3) \qquad \text{Infinite Fourier - Sine Transform}$$

$$\bar{f}(p) = \int_0^\infty f(x).\cos(p\,x)\,dx \qquad (4) \qquad \text{Infinite Fourier - Cosine Transform}$$

$$\overline{f}(p) = \int_0^\infty f(x) . x \, J_0 \, (p \, x) \, dx \qquad (5) \qquad \text{Hankel Transform}$$

$$\overline{f}(p) = \int_0^\infty f(x) . x^{p-1} \, dx \qquad (6) \qquad \text{Mellin Transform}$$

Such transforms have been widely used to obtain solutions of both **ordinary and partial linear differential equations.**

The main object of the present work is to discuss in some detail the Laplace transformation and its applications to the solution of differential equations with given **initial conditions** and in solving **boundary-value problems**.

1.2. Some elementary Laplace transforms

Let F(t) be a function defined for all positive values of the real variable t.
The Laplace transform of F(t) usually denoted by

$$L\{F(t)\} = \int_0^\infty e^{-st} \, F(t) \, dt \ = f(s) \qquad (7)$$

provided that the integral exists.

We shall denote the original function by a capital letter and its transform by the corresponding lower case letter.

We often call F(t) the **object function** and f(s) the **result function**.
The variable s may be real or complex but we shall consider first real values of s.

We shall start by finding the Laplace transforms of some of the **elementary functions** from first principles. Later on, simpler methods will be used for the derivation of these transforms and we shall give the limitations on F(t) as well as on the range of s.

I. Laplace transform of constant: F(t) = 1, where, t > 0

$$L\{1\} = \int_0^\infty e^{-st} \, dt \ = \left[\frac{e^{-st}}{-s} \right]_0^\infty \qquad (8)$$

$$= \frac{1}{s} , \qquad s > 0$$

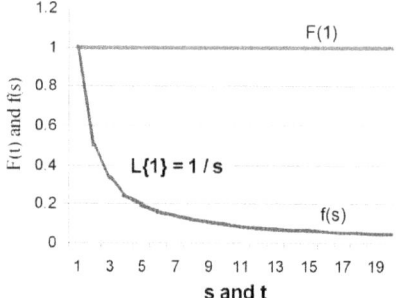

II. Laplace transform of exponential function: $F(t) = e^{at}$, where, $t > 0$

$$L\{e^{at}\} = \int_0^\infty e^{-st}.e^{at} \ dt = \int_0^\infty e^{-(s-a)t} \ dt \qquad (9)$$

$$= -\left[\frac{e^{-(s-a)t}}{s-a}\right]_0^\infty = \frac{1}{s-a} \qquad s > a$$

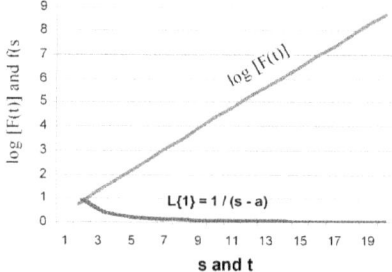

III. Laplace transform of power function: $F(t) = t^k$, where, $k > -1$

$$L\{t^k\} = \int_0^\infty e^{-st}.t^k \ dt \qquad (10)$$

Equation (10) could be converted to the conventional equation of gamma function, specifically

$$\Gamma(n) = \int_0^\infty e^{-x}.x^{n-1} \ dx \qquad (11)$$

In order to convert equation (10) to take the form of (11), we will substitute by

$$x = st, \qquad \text{where, } s > 0$$

where,

4

$$t = \frac{x}{s}, \quad \text{where upon differentiation gives} \quad dt = \frac{dx}{s}$$

Therefore,

$$L\{t^k\} = \int_0^\infty e^{-x} \cdot \frac{x^k}{s^k} \frac{dx}{s} \tag{12}$$

$$= \frac{1}{s^{k+1}} \cdot \int_0^\infty e^{-x} \cdot x^k dx$$

$$= \frac{\Gamma(k+1)}{s^{k+1}}$$

Note that the condition $k > 1$ is necessary for the convergence of the integral.

Corollary 1:

If k is a positive **integer** equal to n, then

$$L\{t^n\} = \frac{\Gamma(n+1)}{s^{n+1}} = \frac{n!}{s^{n+1}}, \quad s > 0 \tag{13}$$

Thus,

$$L\{t\} = \frac{1}{s^2}$$

$$L\{t^2\} = \frac{2}{s^3}$$

$$L\{t^3\} = \frac{3.2}{s^4} = \frac{6}{s^4}$$

......

Corollary 2:

If $k = -\frac{1}{2}$, then

$$L\{t^{-\frac{1}{2}}\} = \frac{\Gamma(\frac{1}{2})}{s^{\frac{1}{2}}} = \sqrt{\frac{\pi}{s}}, \quad s > 0 \tag{14}$$

We prove equation (14) in details as follows.

$$\Gamma(\tfrac{1}{2}) = \int_0^\infty e^{-x} \cdot x^{-\frac{1}{2}} dx \tag{14-1}$$

Now, substitute by $x = y^2$, which upon differentiation gives

$$dx = 2\, y\, dy$$

Which upon substitution in equation (14-1), gives

9

$$\Gamma\left(\tfrac{1}{2}\right) = \int_0^\infty e^{-y^2} . y^{-1} \, 2 \, y \, dy$$

$$= 2 \int_0^\infty e^{-y^2} dy \tag{14-2}$$

The integral in (14-2) can be integrated by converting its coordinate into the polar coordinates as follows

$$I = \int_0^\infty e^{-y^2} dy$$

Therefore, by squaring, we get

$$I^2 = \int_0^\infty \int_0^\infty e^{-y^2} e^{-x^2} dy \, dx \tag{14-3}$$

Substituting by

$$dx \, dy = r \, dr \, d\theta$$
$$r^2 = x^2 + y^2$$

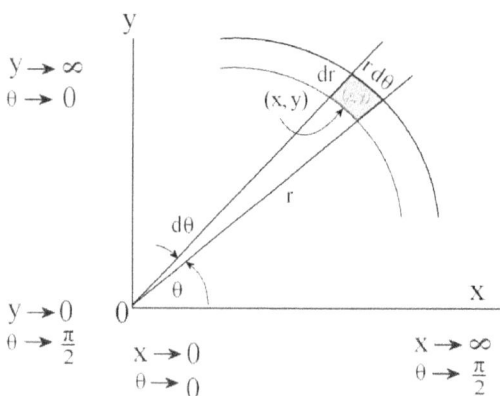

$y \to \infty$
$\theta \to 0$

$y \to 0$
$\theta \to \frac{\pi}{2}$

$x \to 0$
$\theta \to 0$

$x \to \infty$
$\theta \to \frac{\pi}{2}$

We get

$$I^2 = \int_{\theta=0}^{\pi/2} \int_{r=0}^{\infty} e^{-r^2} r \, dr \, d\theta \tag{14-4}$$

$$= \tfrac{1}{2} \int_{\theta=0}^{\pi/2} \int_{r=0}^{\infty} e^{-r^2} dr^2 \, d\theta \tag{14-5}$$

$$= \frac{1}{2} \int_{r=0}^{\infty} \left[\theta \, e^{-r^2} \right]_0^{\pi/2} dr^2 \qquad (14\text{-}6)$$

$$= -\frac{\pi}{4}(0-1) = \frac{\pi}{4} \qquad (14\text{-}7)$$

From (14-2) and (14-7), we get

$$\Gamma\left(\tfrac{1}{2}\right) = 2 \int_0^{\infty} e^{-y^2} . dy = \quad 2.I \qquad (14\text{-}8)$$

$$= \quad 2 \frac{\sqrt{\pi}}{2} \ = \ \sqrt{\pi} \qquad (14\text{-}9)$$

IV. Sine function: F(t) = sin (kt)

$$L\{\sin kt\} = \int_0^{\infty} e^{-st} . \sin(kt)\ dt \qquad (15)$$

The integral in equation (15) requires successive steps of integrating by parts, as follows.

Step 1:

Replace the differential term [sin (kt) **d** t] by [(-1 / k) **d** (cos kt)]. We get

$$L\{\sin kt\} = \int_0^{\infty} e^{-st} . \left[\frac{-1}{k} \right] . d(\cos kt) \qquad (15\text{-}1)$$

Step 2:

First stage in integration-by-parts follows the formula

$$\int u\, dv = uv - \int v\, du\,.$$

In equation (15-1), the u and v are given by

$$u = e^{-st} . \left[\frac{-1}{k} \right] \qquad (15\text{-}1a)$$

$$v = \cos kt \qquad (15\text{-}1b)$$

Thus, equation (15-1) becomes

$$= -\left[\frac{1}{k} \right] \left[e^{-st} . \cos kt) \right]_0^{\infty} + \left[\frac{1}{k} \right] \int_0^{\infty} \cos kt\ d\,(e^{-st}) \qquad (15\text{-}2)$$

Step 3:

11

Arranging the differentials from [d (e⁻ˢᵗ)] to [-s e⁻ˢᵗ dt]. This step prepares to rearrange the cosine term in order to obtain the terms of equation (15) on the right and left sides of the equation, as will be seen in few steps.

$$= -\left[\frac{1}{k}\right]\left[e^{-st}.\cos kt)\right]_0^\infty - \left[\frac{s}{k}\right]\int_0^\infty e^{-st}\cos kt\ dt \qquad (15\text{-}3)$$

Step 4:

Rearranging the differentials from [(cos kt) d t] to [(1/k) d (sin kt)]. This step will prepare us to get the sine term in the form of equation (15) after performing integration-by-parts.

$$= -\left[\frac{1}{k}\right]\left[e^{-st}.\cos kt)\right]_0^\infty - \left[\frac{s}{k^2}\right]\int_0^\infty e^{-st}d(\sin kt) \qquad (15\text{-}4)$$

Step 5:

Last step in integration-by-parts on the integral of equation (15-4) will generate a term equal to that of equation (15), on both sides of the equation (15-4). In equation (15-4), the u and v are given by

$$u = e^{-st}. \qquad (15\text{-}4a)$$

$$v = \sin kt \qquad (15\text{-}4b)$$

Thus, equation (15-4) becomes

$$L\{\sin kt\} = -\left[\frac{1}{k}\right]\left[e^{-st}.\cos kt)\right]_0^\infty - \left[\frac{s}{k^2}\right]\left[e^{-st}\sin kt\right]_0^\infty + \left[\frac{s}{k^2}\right]\int_0^\infty \sin kt\ d\ (e^{-st}) \qquad (15\text{-}5)$$

Step 6:

Repeating Step 3, the differentials from [d (e⁻ˢᵗ)] changes to [-s e⁻ˢᵗ dt]. Thus, equation (15-5) becomes.

$$L\{\sin kt\} = -\left[\frac{1}{k}\right]\left[e^{-st}.\cos kt)\right]_0^\infty - \left[\frac{s}{k^2}\right]\left[e^{-st}\sin kt\right]_0^\infty - \left[\frac{s^2}{k^2}\right]\int_0^\infty e^{-st}\sin kt\ dt \qquad (15\text{-}6)$$

Step 7:

12

Thus, with two steps of integration-by-parts and four steps of rearranging the differentials, we obtained equation (15-6) which contains the term L { sin kt } on both sides of the equation pursuant to equation (15).

Thus, equation (15-6) is written in terms of L { sin kt } as follows

$$L\{\sin kt\} = -\left[\frac{1}{k}\right]\left[e^{-st} \cdot \cos kt)\right]_0^\infty - \left[\frac{s}{k^2}\right]\left[e^{-st}\sin kt\right]_0^\infty - \left[\frac{s^2}{k^2}\right]L\{\sin kt\} \qquad (15\text{-}7)$$

Step 8:

We can now lump the terms L { sin kt } on the left side of equation (15-7) to get:

$$L\{\sin kt\}\left(1+\left[\frac{s^2}{k^2}\right]\right) \quad = \quad -\left[\frac{1}{k}\right]\left[e^{-st} \cdot \cos kt)\right]_0^\infty - \left[\frac{s}{k^2}\right]\left[e^{-st}\sin kt\right]_0^\infty \qquad (15\text{-}8)$$

$$L\{\sin kt\}\left[\frac{k^2+s^2}{k^2}\right] \quad = \quad -\left[\frac{1}{k}\right]\left[e^{-st} \cdot \cos kt)\right]_0^\infty - \left[\frac{s}{k^2}\right]\left[e^{-st}\sin kt\right]_0^\infty \qquad (15\text{-}9)$$

Step 9:

We can now obtain an expression for L { sin kt } on the left side of equation (15-9) devoid of any integral as follows:

$$L\{\sin kt\} = -\left[\frac{k^2}{k^2+s^2}\right]\left[\frac{1}{k}\right]\left[e^{-st} \cdot \cos kt)\right]_0^\infty - \left[\frac{k^2}{k^2+s^2}\right]\left[\frac{s}{k^2}\right]\left[e^{-st}\sin kt\right]_0^\infty \qquad (15\text{-}10)$$

$$L\{\sin kt\} = -\left[\frac{k}{k^2+s^2}\right]\left[e^{-st} \cdot \cos kt)\right]_0^\infty - \left[\frac{s}{k^2+s^2}\right]\left[e^{-st}\sin kt\right]_0^\infty \qquad (15\text{-}11)$$

$$L\{\sin kt\} = -\left[\frac{k}{k^2+s^2}\right]\left[e^{-st} \cdot \cos kt)\right]_0^\infty - \left[\frac{s}{k^2+s^2}\right]\left[e^{-st}\sin kt\right]_0^\infty \qquad (15\text{-}12)$$

Step 10:

The limits of integration are critical to convergence since the term e^{-st} vanishes as $t \to \infty$ and sin (kt) vanishes as $t \to 0$. Thus, equation (15 12) gives

$$L\{\sin kt\} = -\left[\frac{k}{k^2+s^2}\right] - \left[\frac{0}{k^2+s^2}\right] \qquad (15\text{-}13)$$

$$L\{\sin kt\} = \frac{k}{k^2+s^2} \qquad (15\text{-}14)$$

V. Cosine function: F(t) = cos (kt)

13

$$L\{\cos kt\} = \int_0^\infty e^{-st} . \cos(kt)\ dt \qquad (16)$$

The integral in equation (16) can be evaluated in the same manner as above, with successive steps of integrating by parts. That gives

$$L\{\cos kt\} = \left[\frac{k}{k^2 + s^2}\right]\left[e^{-st}.\sin kt)\right]_0^\infty - \left[\frac{s}{k^2 + s^2}\right]\left[e^{-st}\cos kt\right]_0^\infty \qquad (16\text{-}1)$$

$$L\{\cos kt\} = -\left[\frac{0}{k^2 + s^2}\right] - \left[\frac{-s}{k^2 + s^2}\right] \qquad (16\text{-}2)$$

$$L\{\cos kt\} = \frac{s}{k^2 + s^2} \qquad , \ s > 0 \qquad (16\text{-}3)$$

1.3. The Laplace transformation of the sum of two functions

Let A and B be two arbitrary constants, then

$$L\{A F(t) + B G(t)\} = \int_0^\infty e^{-st} \{A F(t) + B G(t)\} dt$$

$$= A \int_0^\infty e^{-st} F(t) dt + B \int_0^\infty e^{-st} G(t) dt$$

$$= A L\{F(t)\} + B L\{G(t)\}$$

$$= A f(s) + B g(s) \qquad (17)$$

This means that the Laplace transform of a linear combination of two functions is the same linear combination of the transforms of these functions.

Example 1: Hyperbolic cosh function

$$L\{\cosh kt\} = L\{\frac{e^{kt} + e^{-kt}}{2}\}$$

$$= 2\left[\frac{1}{s-k} + \frac{1}{s+k}\right]$$

$$= \frac{s}{s^2 - k^2}, \qquad s > |k| \qquad (18)$$

Similarly,

$$L\{\sinh kt\} = \frac{k}{s^2 - k^2}, \qquad s > |k| \qquad (19)$$

Table 2. Laplace Transforms of elementary functions

F(t)	L{(t)}	f(s)	
1	L{1}	$\dfrac{1}{s}$,	$s > 0$
e^{at}	L{e^{at}}	$\dfrac{1}{s-a}$	$s > a$
t^n	L{t^n}	$\dfrac{\Gamma(n+1)}{s^{n+1}} = \dfrac{n!}{s^{n+1}}$,	$s > 0$
$t^{-\frac{1}{2}}$	L{$t^{-\frac{1}{2}}$}	$\sqrt{\dfrac{\pi}{s}}$, $s > 0$	
$\sin kt$	L{$\sin kt$}	$\dfrac{k}{k^2 + s^2}$	
$\cos kt$	L{$\cos kt$}	$\dfrac{s}{k^2 + s^2}$	, $s > 0$
$\cosh kt$	L{$\cosh kt$}	$\dfrac{s}{s^2 - k^2}$,	$s > |k|$
$\sinh kt$	L{$\sinh kt$}	$\dfrac{k}{s^2 - k^2}$,	$s > |k|$

1.4. Sectionally or piecewise continuous functions

A function is said to be sectionally **continuous or piecewise continuous** in an interval
$a \le t \le b$ if this interval can be subdivided into a finite number of intervals in each of which the
function is continuous and has **finite right and left hand limits**. It follows from the definition
that sectionally continuous functions include two main classes of functions:

(i) Continuous functions
(ii) Discontinuous functions in which the discontinuity consists of finite jumps

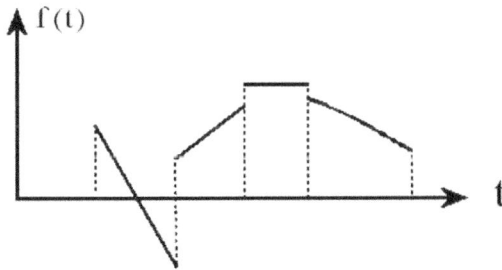

It is worthy of notice that sectionally continuous functions are integrable functions, the integral
being the **sum of the integrals** of the continuous functions over the subintervals.

15

Example 2 : Step function

Find the Laplace transform of the function F(t) defined by

$F(t) = F_0 \quad 0 < t < t_0$
$F(t) = 0 \qquad t > t_0$

Solution

$$L\left\{F(t)\right\} = \int\limits_0^\infty e^{-st} \, F(t) \, dt$$

$$= \int\limits_0^{t_0} e^{-st} \, F_0 \, dt$$

$$= F_0 \left[\frac{e^{-st}}{-s}\right]_0^{t_0}$$

$$= \frac{F_0}{s}\left(1 - e^{-st_0}\right)$$

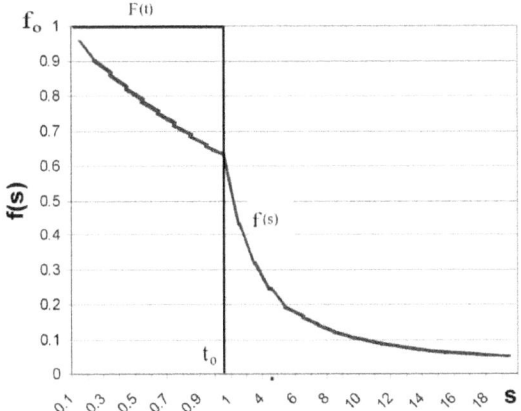

Example 3: Discontinuous fucntion

Find the Laplace transform of H(t), where

$H(t) = t$ at $0 < t < 4$ and $H(t) = 5$ at $t > 4$

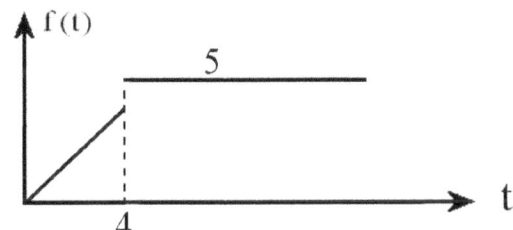

Solution:

$$L\{H(t)\} = \int_0^\infty e^{-st} H(t)\, dt$$

$$= \int_0^4 e^{-st} \cdot t \cdot dt + \int_4^\infty e^{-st} \cdot 5 \cdot dt$$

$$= \left[-\frac{t}{s} e^{-st} - \frac{t}{s^2} e^{-st} \right]_0^4 + \left[-\frac{5}{s} e^{-st} \right]_4^\infty$$

$$= \frac{1}{s^2} + \frac{e^{-4s}}{s} - \frac{e^{-4s}}{s^2}$$

17

1.5. Functions of exponential order

If two positive constants M and a exist such that for all $t > T$ the following inequality holds true

$$|e^{-at} F(t)| < M \quad \text{Or} \quad |F(t)| < M e^{at}$$

Then, we say that $F(t)$ is a function of exponential order a as $t \to \infty$ and write

$F(t)$ is $O(e^{at})$.

This actually means that functions of exponential order cannot grow in absolute value more rapidly than $M e^{at}$ as t increases.

Example 4

Let $F(t) = C$ where C is a constant. Choose a positive constant M such that $M > |C|$

Therefore, $|C| < M e^{at}$ as $t \to \infty$ for all $a > 0$
Therefore, C is $O(e^{at})$ where $a > 0$

Similarly for all bounded functions e.g. **sin (kt)** or **cos (kt).** Thus **sin (kt)** and **cos (kt)** are $O(e^{at})$ where $a > 0$

Example 5

Let $F(t) = t^n$ where n is a positive integer. Therefore, the integrand of Laplace integral

$$\frac{t^n}{e^{at}} \to 0 \text{ as } t \to \infty \text{ for all } a > 0$$

Therefore,

$$\left| \frac{t^n}{e^{at}} - 0 \right| < M \text{ as } t \to \infty$$

i.e., $|t^n| < M e^{at}$ as $t \to \infty$ for all $a > 0$

Therefore,

t^n is $O(e^{at})$ where $a > 0$.

Example 6

Let $F(t) = e^{t^2}$

Therefore, the Laplace integrand is

$$\left| e^{-\alpha t} \cdot e^{t^3} \right| = e^{-\alpha t + t^3}$$

can be made larger than any given constant by increasing t. Hence e^{t^3} **is NOT** of exponential order.

1.6. Sufficient conditions for the existence of the Laplace transform

We shall now show that the Laplace transform of F(t) exists when:

(i)   F(t) is sectionally continuous in every finite interval in the range $t \geq 0$.
(ii)   F(t) is $\mathbf{O}(e^{\alpha t})$ where $t \to \infty$

i.e., $|F(t)| < M e^{\alpha t}$ for all $t > T$ (arbitrary constant), such that the Laplace transform becomes

$$\int_0^\infty e^{-st} F(t) \, dt = \int_0^T e^{-st} F(t) \, dt + \int_T^\infty e^{-st} F(t) \, dt$$

The first integral on the R.H.S. exists since F(t) is sectionally continuous in every finite interval $0 \leq t \leq T$.
The second integral on the right also exists since F(t) is $\mathbf{O}(e^{\alpha t})$ for $t > T$. This follows from the fact that

$$\left| \int_T^\infty e^{-st} F(t) \, dt \right| \leq \int_T^\infty \left| e^{-st} F(t) \right| \, dt < \int_T^\infty e^{-st} M e^{\alpha t} \, dt$$

$$= M \int_T^\infty e^{-(s-\alpha)t} \, dt = M \left[\frac{e^{-(s-\alpha)t}}{-(s-\alpha)} \right]_T^\infty$$

$$= M \frac{e^{-(s-\alpha)T}}{(s-\alpha)}, \qquad s > \alpha$$

$$< \frac{M}{s-\alpha}$$

Hence, the Laplace integral $\int_0^\infty e^{-st} F(t) \, dt$ exists if $s > a$.

It should be noticed that the above conditions though **sufficient** are **not necessary**. They are however simple to apply in most practical problems. If the above conditions are not satisfied the Laplace transform may or may not exist. Thus for example

$$L\{t^{-\frac{1}{2}}\} = \sqrt{\frac{\pi}{s}}, \qquad s > 0$$

as we have already shown in section (1.2), but $F(t) = t^{-\frac{1}{2}}$ does not satisfy the above conditions.

19

1.7. Null functions

A null function N(t) is a function of t such that for all t > 0 the following condition holds true

$$\int_0^t N(\tau)\,d\tau = 0 \qquad\qquad (20)$$

Example 7

The function defined by the following points, and shown in the figure below,

F(t) = 2, t = 1
F(t) = 1, t = 2
F(t) = -1, t = 3
F(t) = 0 otherwise

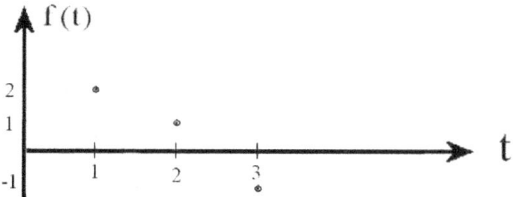

is a null function. In general, any function, which is zero at all but a countable set of **points** is a null function. It is evident that the Laplace transform of a null function is **zero**. It should be noticed that having values at certain points, **without being continuous between any of those points**, renders the integral null.

1.8. Inverse Laplace transforms

If   L { F(t)} = f(s), then the inverse Laplace transform is often written

$$L^{-1}\{f(s)\} = F(t) \qquad\qquad (21)$$

i.e., the inverse Laplace transform of f(s) is F(t).

Now, the Laplace transform of a null function N (t) is zero.
Hence if

$$L\{F(t)\} = f(s) \text{ then } L\{F(t) + N(t)\} = f(s) \qquad\qquad (22)$$

Thus, we can have **two different functions with the same Laplace transform**.

Hence, if we allow **null functions**, the inverse transform is **not unique**, but if we do not allow null functions, then it has been shown [by Mathias Lerch] that if we restrict ourselves to functions of exponential order for $t > T$, and which are sectionally continuous in every finite interval $0 \leq t \leq T$, then, the inverse Laplace transform equation (21) is unique.

We shall now write the previous transforms which we have already obtained in the inverse form.

$$L^{-1}\left\{\frac{1}{s}\right\} = 1 \tag{23-1}$$

$$L^{-1}\left\{\frac{1}{s-a}\right\} = e^{\alpha t} \tag{23-2}$$

$$L^{-1}\left\{\frac{1}{s+a}\right\} = e^{-\alpha t} \tag{23-3}$$

$$L^{-1}\left\{\frac{1}{s^n}\right\} = \frac{t^{n-1}}{(n-1)!}, \quad (n \text{ is a positive integer}) \tag{23-4}$$

$$L^{-1}\left\{\frac{1}{k^2+s^2}\right\} = \frac{1}{k}\sin kt \tag{23-5}$$

$$L^{-1}\left\{\frac{s}{k^2+s^2}\right\} = \cos kt \tag{23-6}$$

$$L^{-1}\left\{\frac{s}{s^2-k^2}\right\} = \cosh kt \tag{23-7}$$

$$L^{-1}\left\{\frac{1}{s^2-k^2}\right\} = \frac{1}{k}\sinh kt \tag{23-8}$$

1.9. The inverse Laplace transformation of the sum of two functions

We have shown by equation (17) that

$$L\left\{A\ F(t)\ +B\ G(t)\right\}\ =\ A\ f(s)\ +B\ g(s)$$

Therefore, the inverse Laplace transform of equation (17) proves that:

21

"The inverse Laplace transform of the **sum of two functions** is the **sum of the inverse Laplace transforms** of the each of two functions."

Thus,

$$L^{-1}\{A\ f(s)\ +B\ g(s)\}\ =\ A\ F(t)\ +B\ G(t)$$
$$=\ A\,L^{-1}\ \{f(s)\}+B\ L^{-1}\{g(s)\}$$

Example 8

Find the inverse Laplace transforms of the expression

$$\frac{3}{s+1}+\frac{2s}{s^2+25}-\frac{4}{s^2+9}$$

Solution

Equations (23) give the inverse Laplace Transform for each term in the above expression as follows.

Put a = 1 in equation (23-3), we get $L^{-1}\left\{\dfrac{3}{s+1}\right\}\ =3\,L^{-1}\left\{\dfrac{1}{s+1}\right\}=\ 3\,e^{-t}$

Put k = 5 in equation (23-6), we get $L^{-1}\left\{\dfrac{2s}{25+s^2}\right\}=2\,L^{-1}\left\{\dfrac{s}{25+s^2}\right\}=2\cos 5t$

Put k = 3 in equation (23-5), we get $L^{-1}\left\{\dfrac{4}{9+s^2}\right\}=4\,L^{-1}\left\{\dfrac{1}{9+s^2}\right\}=4\left(\dfrac{1}{3}\right)\sin 3t$

Next, we use the additive property of inverse Laplace transform to get

$$L^{-1}\left\{\frac{3}{s+1}+\frac{2s}{s^2+25}-\frac{4}{s^2+9}\right\}=3\,e^{-t}+2\cos 5t-\frac{4}{3}\sin 3t$$

1.10. Laplace transforms of derivatives

In section 1.6, we defined the existence of a function, say F (t), to be of $O(e^{at})$ where t → ∞ and that F(t) is continuous with a sectionally continuous derivatives F'(t) in every finite interval 0 ≤ t ≤ T.

Then, provided s > a, we will prove that the Laplace transform of F'(t) exists and is given by

$$L\{F'(t)\}=s\,L\{F(t)\}-F(0)\ =s\ f(s)\ -F(0) \tag{24}$$

Where, owing to the **continuity** of F(t), therefore F(0) is the same as F(0+). i.e., the value of the function when approached from the positive direction of t is the same as the limiting value of F(t) when t approaches 0 through positive values of t.

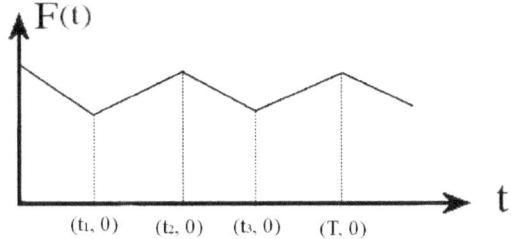

The figure shows a continuous function with a sectionally continuous derivative in the interval ($0 \leq t \leq T$).

To prove the validity of equation (24), we start with the definition of the Laplace transform then we introduce the definition of the differential derivative as follows.

$$L\{F'(t)\} = \lim_{T \to \infty} \int_0^T e^{-st} F'(t) \, dt \tag{24-1}$$

Then, we use the additive property of Laplace transform such that

$$\int_0^T e^{-st} F'(t) \, dt = \int_0^{t_1} e^{-st} F'(t) \, dt + \int_{t_1}^{t_2} e^{-st} F'(t) \, dt + \dots + \int_{t_n}^T e^{-st} F'(t) \, dt \tag{24-2}$$

Equation (24-2) **raises concern on the rationale of adding derivatives** in the same manner we added the original function over the various segments. That is because derivatives are ratios, not absolute functions. Yet, the integration process over the width of each segment renders the ratios absolute values by summing all ratios on the t-axis as follows.

Let us substitute by

$$F'(t) \, dt = \frac{dF(t)}{dt} . dt = dF(t) \tag{24-3}$$

Thus, equation (24-2) becomes

$$\int_0^T e^{-st} F'(t) \, dt = \int_0^{t_1} e^{-st} dF(t) + \int_{t_1}^{t_2} e^{-st} dF(t) + \dots + \int_{t_n}^T e^{-st} dF \, dt \tag{24-4}$$

Integrating-by-parts we using the formula $\int u \, dv = uv - \int v \, du$, equation (24-4) becomes

$$\int_0^T e^{-st} F'(t) \, dt = \left[e^{-st} F(t) \right]_0^{t_1} + \left[e^{-st} F(t) \right]_{t_1}^{t_2} + \dots + \left[e^{-st} F(t) \right]_{t_n}^T$$

23

$$- \int_0^{t_1} F(t)\ de^{-st} - \int_{t_1}^{t_2} F(t)\ de^{-st} \quad ..... - \int_{t_n}^{T} F(t)\ de^{-st} \tag{24-5}$$

Equation (24-5) is expanded using the two properties

The unity property: $e^{-0s} = 1$

and

The differential property: $(d\ e^{-st}) = (-s\ e^{-st}\ dt)$

Thus, equation (24-5) becomes

$$\int_0^{T} e^{-st}\ F'(t)\ dt = \left[e^{-st_1}\ F(t_1) - F(0) \right] + \left[e^{-st_2}\ F(t_2) - e^{-st_1}\ F(t_1) \right] + ... + \left[e^{-sT}\ F(T) - e^{-st_n}\ F(t_n) \right]$$

$$- (-s) \left[\int_0^{t_1} e^{-st}\ F(t)\ dt + \int_{t_1}^{t_2} e^{-st}\ F(t)\ dt\ ..... + \int_{t_n}^{T} e^{-st}\ F(t)\ dt \right] \tag{24-6}$$

Now, owing to the **continuity** of F(t), therefore

$F(t_{1+}) = F(t_{1-}),\ \ F(t_{2+}) = F(t_{2-}),\ ..,\ F(t_{n+}) = F(t_{n-}),$

The continuity property unites the integrals over the n-segments in equation (24-6). Farther, the equal constants with opposite signs cancel each other leaving only the term [e^{-sT} F(T) – F(0)].

Thus, equation (24-6) becomes

$$\int_0^{T} e^{-st}\ F'(t)\ dt = e^{-sT}\ F(T) - F(0) + s \int_0^{T} e^{-st}\ F(t)\ dt \tag{24-7}$$

Needless to say that the integration over the interval (0 , T) is equivalent to the integration over $(0, \infty)$ since F(t) vanishes at $t > T$.

Now, since F(t) is O (e^{at}) , then $| F(t) | < M\ e^{at}$ for large t and consequently

$$| e^{-sT}\ F(t) |\ < M\ e^{-(s-a)\ T} \rightarrow o\ as\ T \rightarrow \infty \quad for\ s > a$$

Also

$$\int_0^{T} e^{-st}\ F(t)\ dt\ \rightarrow L\left\{ F(t) \right\}\ as\ T \rightarrow \infty \tag{24-8}$$

And

$$e^{-sT}\ F(T)\ \rightarrow 0\ as\ T \rightarrow \infty \tag{24-9}$$

Therefore

$$\int_0^\infty e^{-st} F'(t)\ dt = -F(0) + s\ L\{F(t)\} \qquad (24\text{-}10)$$

i.e.,

$$L\{F'(t)\} = s\ f(s) - F(0) \qquad (24\text{-}11)$$

Example 9

Given the function

$$F(t) = t + 1, \quad 0 \le t \le 2$$
$$= 3, \qquad\qquad t > 2$$

Draw the graphs of F(t) and F'(t) and find L { F' (t) }

Solution

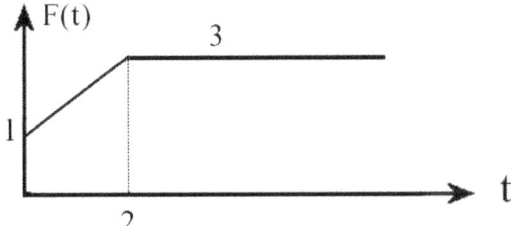

$$F(0) = 1 \qquad (24\text{-}12)$$

$$f(s) = \int_0^2 e^{-st}(t+1)\ dt + 3\int_2^\infty e^{-st}\ dt$$

$$= \int_0^2 \frac{t}{-s}\ d\,e^{-st} + \int_0^2 e^{-st}\ dt + 3\int_2^\infty e^{-st}\ dt$$

$$= \left[\frac{te^{-st}}{-s}\right]_0^2 - \int_0^2 \frac{e^{-st}}{-s}\ dt + \left[\frac{e^{-st}}{-s}\right]_0^2 + 3\left[\frac{e^{-st}}{-s}\right]_2^\infty$$

$$= \left[\frac{2e^{-2s}}{-s}\right] - \left[\frac{e^{-st}}{s^2}\right]_0^2 + \left[\frac{e^{-2s}-1}{-s}\right] + 3\left[\frac{-e^{-2s}}{-s}\right]$$

$$= \left[\frac{2e^{-2s}}{-s}\right] - \left[\frac{e^{-2s}-1}{s^2}\right] + \left[\frac{e^{-2s}-1}{-s}\right] + 3\left[\frac{-e^{-2s}}{-s}\right]$$

$$= \frac{1}{s}\left[1 - \frac{e^{-2s}-1}{s}\right] \qquad (24\text{-}13)$$

Substituting by (24-12) and (24-13) in Equation (24) we get

25

$$L\{F'(t)\} = s\ f(s) - F(0)$$

$$= s\left(\frac{1}{s}\left[1 - \frac{e^{-2s}-1}{s}\right]\right) - 1$$

$$= -\frac{e^{-2s}-1}{s}$$

$$= \frac{1}{s}\left(1 - e^{-2s}\right),\ \text{for}\ s > 0$$

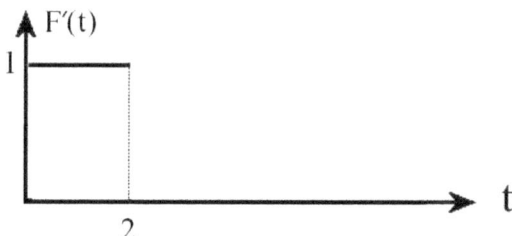

The transform of the **second derivative** F"(t) can be obtained in a similar manner by applying the above theorem to F'(t). Assuming F(t) to be continuous and of exponential order and F"(t) to be sectionally continuous, then

$$L\{F''(t)\} = s\ L\{F'(t)\} - F'(0)$$
$$= s\ [s\ f(s) - F(0)] - F'(0)$$
$$= s^2\ f(s) - s\ F(0) - F'(0) \qquad\qquad (25)$$

Example 10

Find L { sinh kt} by using the Laplace transform of derivatives of functions, equation (25).

Solution

Put

F(t) = sinh kt

Then,

F'(t) = k cosh kt
F''(t) = k^2 sinh kt
F(0)   = 0
F'(0) = k

26

Substituting with the above expressions in equation (25) gives

$$L\{F''(t)\} = s^2 f(s) - s F(0) - F'(0)$$
$$k^2 L\{\sinh kt\} = s^2 L\{\sinh kt\} - 0 - k \qquad (25\text{-}1)$$

Arranging the terms in the above equation we get

$$(k^2 - s^2) L\{\sinh kt\} = -k$$

Or

$$L\{\sinh kt\} = \frac{k}{(s^2 - k^2)} \qquad (25\text{-}2)$$

We have proved equation (25-2) also in equation (19)

In general, suppose that F(t) together with its derivatives

$$F'(t), F''(t),.., F^{(n-1)}(t)$$

to be of order $O(e^{at})$ as $t \to \infty$ and that F(t) has a **continuous** derivative $F^{(n-1)}(t)$ and a sectionally continuous derivative $F^{(n)}(t)$ in every finite interval ($0 \le t \le T$), then

$$L\{F^{(n)}(t)\} = s^n f(s) - s^{n-1} F(0) - s^{n-2} F'(0) - ... F^{(n-1)}(0) \qquad (26)$$

Example 11

Find $L\{t^n\}$ where n is a positive integer.

Solution

Put

$$F(t) = t^n$$

Then

$$F'(t) = n t^{n-1}$$
$$F''(t) = n (n-1) t^{n-2}$$
$$\cdots$$
$$F^n(t) = n!$$

Therefore,

$$F'(0) = F''(0) = ... = F^n(0) = 0$$

Substituting with the above expressions in equation (26) gives

$$L\left\{F^{(n)}(t)\right\}= s^n\ f(s)- s^{n-1}\ F(0)- s^{n-2}\ F'(0)-...F^{(n-1)}(0)$$
$$L\{n!\}= s^n\ L\{t^n\}-0\ -0\ ....\qquad\qquad (26\text{-}1)$$

But, we also proved that

$$L\{1\}= \frac{1}{s}$$

Therefore,

$$L\{n!\}= \frac{n!}{s}\qquad\qquad (26\text{-}2)$$

Thus, from (26-1) and (26-2), we get

$$L\{n!\}= \frac{n!}{s} = s^n\ L\{t^n\}$$

Therefore,

$$L\{t^n\}= \frac{n!}{s^{n+1}}\qquad\qquad (26\text{-}3)$$

Equation (26-3) is the same as equation (13).

1.11. Laplace transforms of integrals.

If F(t) be sectionally **continuous** and of **exponential order** then

$$L\left\{ \int_0^t F(\tau)\,d\tau \right\}= \frac{1}{s}\ f(s)\qquad\qquad (27)$$

Where,

$$f(s) = L\ \{\ F(t)\qquad\qquad (27\text{-}1)$$

Put

$$G(t)= \int_0^t F(\tau)\,d\tau\qquad\qquad (27\text{-}2)$$

Then, G(t) is continuous and of exponential order and G(0) = 0.
Since,

$$G'(t)\ = F(t)\qquad\qquad (27\text{-}3)$$

Then, transforming both sides of (27-3) and using the property of Laplace transforming of derivatives we get

$$L\{G'(t)\}= s\,L\{G(t)\}-G(0)$$
$$= L\{F(t)\}\ =\ f(s)\qquad\qquad (27\text{-}4)$$

28

Therefore,

$$L\{G(t)\} = \frac{1}{s}\left(f(s) + G(0) \right) \tag{27-5}$$

Hence, we have the following result:

"The algebraic **division** of Laplace transform of a function by s corresponds to the **integration** of that function between 0 and t."

The rule can be extended to a **repeated division** by s.
Thus, the division of the transform of F(t) by s^n corresponds to a **repeated integration** of F(t); n times from 0 to t.

Example 12

We know that

$$L\{\sin kt\} = \frac{k}{s^2 + k^2}, \qquad for \quad s > 0$$

Therefore,

$$L\left\{ \int_0^t \sin k\tau \, d\tau \right\} = \frac{1}{s} \cdot \frac{k}{s^2 + k^2} \tag{27-6}$$

$$L\left\{ \left[-\frac{\cos k\tau}{k} \right]_0^t \right\} = \frac{1}{s} \cdot \frac{k}{s^2 + k^2}$$

i.e., $\qquad L\left\{ \frac{1}{k}\left(1 - \cos kt \right) \right\} = \frac{k}{s\left(s^2 + k^2 \right)} \tag{27-7}$

Integrating both sides of equation (27-7) again, we get:

$$L\left\{ \frac{1}{k}\left(t - \frac{1}{k}\sin kt \right) \right\} = \frac{k}{s^2\left(s^2 + k^2 \right)} \tag{27-8}$$

Hence, we have the following two results:

$$L^{-1}\left\{ \frac{1}{s\left(s^2 + k^2 \right)} \right\} = \frac{1}{k^2}\left(1 - \cos kt \right)$$

$$L^{-1}\left\{ \frac{1}{s^2\left(s^2 + k^2 \right)} \right\} = \frac{1}{k^3}\left(kt - \sin kt \right)$$

1.12 The first shift theorem of multiplying the object function by e^{at}

Let F(t) be a sectionally continuous function of order e^{at}. Then

$$L\left\{F(t)\right\} = \int_0^\infty e^{-st} F(t)\, dt = f(s), \quad s > \alpha$$

Therefore,

$$L\left\{e^{at} F(t)\right\} = \int_0^\infty e^{-st} . e^{at} F(t)\, dt$$

$$= \int_0^\infty e^{-(s-a)t} F(t)\, dt$$

$$= \int_0^\infty e^{-(s-a)t} F(t)\, dt$$

$$= f(s-a) \tag{28}$$

Where, $s - a > \alpha$, i.e. $s > a + \alpha$

Hence, we have the following theorem on substitution:

"The change of the variable s, in the transform f(s) of F(t) into s – a, corresponds to the multiplication of F(t) by e^{at}"

Since f(s-a) is a shift of f(s) to the right a distance a, the above theorem is known as the "**first shift theorem**".

Example 13

$$L\left\{e^{at} \sin kt\right\} = \frac{k}{(s-a)^2 + k^2}, \quad s > a$$

$$L\left\{e^{-at} \cos kt\right\} = \frac{s+a}{(s+a)^2 + k^2}, \quad s > -a$$

$$L^{-1}\left\{\frac{1}{(s-a)^n}\right\} = e^{at} \frac{t^{n-1}}{(n-1)!}$$

$$L\left\{e^{3t} \cosh 4t\right\} = \frac{s-3}{(s-3)^2 - 16} = \frac{s-3}{s^2 - 6s - 7}$$

Example 14: Shift function of Laplace transform

Prove that

$$L\left\{e^{at} t^{-\frac{1}{2}} (1 + 2at)\right\} = \frac{\sqrt{\pi}\ s}{(s-a)^{\frac{3}{2}}} \tag{29}$$

Solution

30

We will use equation (13) which gives

$$L\{t^k\} = \frac{\Gamma(k+1)}{s^{k+1}}, \quad k > -1 \tag{29-1}$$

Then, we will consider each term on the LHS of equation (29) as follows

First term:

We will first use the expression

$$L\{t^{-\frac{1}{2}}\} = \sqrt{\frac{\pi}{s}}$$

Then we will use the first-shift theorem to get

$$L\{e^{at} t^{-\frac{1}{2}}\} = \sqrt{\frac{\pi}{s-a}} \tag{29-2}$$

Second term:

We will first use the expression

$$L\{t^{\frac{1}{2}}\} = \frac{\Gamma(\frac{3}{2})}{s^{\frac{3}{2}}} = \frac{\frac{1}{2}\Gamma(\frac{1}{2})}{s^{\frac{3}{2}}} = \frac{\sqrt{\pi}}{2s^{\frac{3}{2}}}$$

Then we will use the first-shift theorem, equation (28), to get

$$L\{e^{at} t^{\frac{1}{2}}\} = \frac{\sqrt{\pi}}{2(s-a)^{\frac{3}{2}}} \tag{29-3}$$

Substituting from (29-2) and (29-3) in (29), we get

$$L\{e^{at} t^{-\frac{1}{2}}(1+2at)\} = L\{e^{at} t^{-\frac{1}{2}}\} + 2a L\{e^{at} t^{\frac{1}{2}}\}$$

$$= \sqrt{\frac{\pi}{s-a}} + 2a \frac{\sqrt{\pi}}{2(s-a)^{\frac{3}{2}}}$$

$$= \sqrt{\frac{\pi}{s-a}}\left(1 + \frac{a}{(s-a)}\right) = \sqrt{\frac{\pi}{s-a}}\left(\frac{s}{(s-a)}\right)$$

1.13. Two useful Laplace transforms: Laplace transforms of (x sin x) and (x cos x)

(i) Given that

$$L\{\cos kt\} = \int_0^\infty e^{-st} \cos kt \, dt = \frac{s}{s^2+k^2}, \quad s > 0 \tag{30}$$

31

Differentiate both sides of (30) with respect to k, we get

$$\frac{\partial}{\partial k}\left[\int_0^\infty e^{-st}\cos kt\, dt\right] = \int_0^\infty \frac{\partial}{\partial k}\left[e^{-st}\cos kt\right] dt$$

$$= -\int_0^\infty e^{-st}\, t\, \sin kt \quad dt$$

$$= -L\{t\sin kt\}$$

$$= -\frac{2ks}{\left(s^2+k^2\right)^2} \quad , \quad s>0 \tag{30-1}$$

Therefore,

$$\boxed{L\{t\sin kt\} = \frac{2ks}{\left(s^2+k^2\right)^2}} \tag{31-1}$$

$$\boxed{L^{-1}\left\{\frac{s}{\left(s^2+k^2\right)^2}\right\} = \frac{1}{2k}\,t\sin kt} \tag{31-2}$$

(ii) Dividing the transform in (31-2) by s, which corresponds to the integration of the object function between 0 and t, we get

$$L^{-1}\left\{\frac{1}{\left(s^2+k^2\right)^2}\right\} = \frac{1}{2k}\int_0^t \tau\sin k\tau\, d\tau \tag{32}$$

We will arrange the differential from ($\sin k\tau\, d\tau$) to ($-1/k$) $d(\cos k\tau)$ to get

$$L^{-1}\left\{\frac{1}{\left(s^2+k^2\right)^2}\right\} = -\frac{1}{2k^2}\int_0^t \tau\, d(\cos k\tau)$$

Integrating-by-parts we get

$$L^{-1}\left\{\frac{1}{\left(s^2+k^2\right)^2}\right\} = -\frac{1}{2k^2}\left[\tau\cos k\tau - \frac{\sin k\tau}{k}\right]_0^t$$

$$= -\frac{1}{2k^2}\left(t\cos kt - \frac{\sin kt}{k}\right) \tag{32-1}$$

Also

$$L\left\{t\cos kt - \frac{\sin kt}{k}\right\} = -\frac{2k^2}{\left(s^2+k^2\right)^2}$$

$$L\{t\cos kt\} = L\left\{\frac{\sin kt}{k}\right\} - \frac{2k^2}{\left(s^2+k^2\right)^2}$$

$$= \frac{1}{k} \frac{k}{\left(s^2+k^2\right)} - \frac{2k^2}{\left(s^2+k^2\right)^2}$$

$$= \frac{s^2+k^2-2k^2}{\left(s^2+k^2\right)^2} = \frac{s^2-k^2}{\left(s^2+k^2\right)^2} \qquad (32\text{-}2)$$

Thus,

$$L^{-1}\left\{\frac{1}{\left(s^2+k^2\right)^2}\right\} = -\frac{1}{2k^2}\left(t\cos kt - \frac{\sin kt}{k}\right) \qquad (32\text{-}1)$$

$$L\left\{t\cos kt\right\} = \frac{s^2-k^2}{\left(s^2+k^2\right)^2} \qquad (32\text{-}2)$$

1.14. The multiplication of the variable by a positive constant

Suppose that L { F(t) } = f (s) , for s > a , and let a be a positive constant, then

$$L\left\{F(\alpha t)\right\} = \int_0^\infty e^{-st}\, F(\alpha t)\, dt \qquad (33)$$

Substitute by $\tau = \alpha t$ and $dt = (1/\alpha)\, d\,\tau$.
Then, equation (33) becomes

$$L\left\{F(\alpha t)\right\} = \frac{1}{\alpha}\int_0^\infty e^{-\frac{s\tau}{\alpha}}\, F(\tau)\, d\tau$$

$$= \frac{1}{\alpha}\, f\left(\frac{s}{\alpha}\right), \qquad \frac{s}{\alpha} > a$$

Therefore,

$$L\left\{F(\alpha t)\right\} = \frac{1}{\alpha}\, f\left(\frac{s}{\alpha}\right) \qquad (33\text{-}1)$$

Where, s > aα

Formula (33-1) can also be written in the form

$$L^{-1}\left\{f(c\, s)\right\} = \frac{1}{c}\, F\left(\frac{t}{c}\right), \qquad c > 0 \qquad (33\text{-}2)$$

Example 15

$$L\left\{\sin t\right\} = \frac{1}{s^2+1}$$

33

Therefore,

$$L\{\sin kt\} = \left(\frac{1}{k}\right)\frac{1}{\left(\dfrac{s}{k}\right)^2 + 1} = \frac{k}{s^2 + k^2}, \quad s > 0$$

1.15. Determination of the inverse Laplace transforms by the aid of partial fractions

A proper fraction of the form $\dfrac{p(s)}{q(s)}$, in which the degree of the numerator p(s) is less than that of the denominator q(s), can be resolved in a series of partial fractions.

Let $q(s) = (s - a)(s - b)^r (s^2 + cs + d) (s^2 + gs + m)^n$

We could write the proper fraction $\dfrac{p(s)}{q(s)}$ in terms of its partial fractions as follows

$$\frac{p(s)}{q(s)} = \frac{A}{s-a} + \left[\frac{B_1}{(s-b)^r} + \frac{B_2}{(s-b)^{r-1}} + ... + \frac{B_r}{(s-b)}\right]$$
$$+ \frac{Cs+D}{s^2+cs+d} + \left[\frac{E_1 s + F_1}{(s^2+gs+m)^n} + \frac{E_2 s + F_2}{(s^2+gs+m)^{n-1}} + ... + \frac{E_n s + F_n}{(s^2+gs+m)}\right] \qquad (34)$$

On clearing the fractions on both sides of equation (34), we obtain an identity from which the unknown constants A, B, C, D, and F can be obtained either by substituting the **zeros** of the denominator in the identity or equating **equal powers** on both sides.

The following are examples on finding the inverse Laplace transform by resolving the transform into its partial fractions.

Example 16

Find the inverse Laplace transform of the expression

$$\frac{s^2 + 3}{(s+1)(s-2)(s-4)} \qquad (34\text{-}1)$$

Solution

The proper fraction in (34-1) can be written in terms of its partial fractions as follows

$$\frac{s^2 + 3}{(s+1)(s-2)(s-4)} = \frac{A}{(s+1)} + \frac{B}{(s-2)} + \frac{C}{(s-4)} \qquad (34\text{-}2)$$

In order to determine the constants A, B, and C, we will cross multiply the opposite ends of (34-2) to get.

$$s^2 + 3 = A(s-2)(s-4) + B(s+1)(s-4) + C(s+1)(s-2) \qquad (34\text{-}3)$$

Equating the coefficients of s^2 in equation (34-3), we get:
$$1 = A + B + C \qquad (34\text{-}3a)$$

Equating the coefficients of s in equation (34-3), we get:

$$0 = -6A - 3B - C \qquad (34\text{-}3b)$$

Equating the constants in equation (34-3), we get:

$$3 = 8A - 4B - 2C \qquad (34\text{-}3c)$$

Adding equations (34-3a) and (34-3b) gives

$$1 = -5A - 2B \qquad (34\text{-}3c)$$

Subtracting twice the equation (34-3b) from (34-3c) gives

$$3 = 20A + 2B \qquad (34\text{-}3d)$$

Adding equations (34-3c) and (34-3d) gives

$$A = 4/15 \qquad (34\text{-}3e)$$

Substituting A from equation (34-3e) into (34-3d) gives

$$B = -7/6 \qquad (34\text{-}3f)$$

Substituting A from equation (34-3e) and B from (34-3f) into (34-3b) gives

$$C = 19/10 \qquad (34\text{-}3g)$$

Having determined A, B, and C in equations (34-3e), (34-3f) and (34-3g), equation (34-2) becomes

$$\frac{s^2 + 3}{(s+1)(s-2)(s-4)} = \frac{4}{15(s+1)} - \frac{7}{6(s-2)} + \frac{19}{10(s-4)} \qquad (35)$$

Hence, the inverse Laplace transform of equation (35) is obtained by using the additive property of inverse Laplace transform as follows

35

$$L^{-1}\left\{\frac{s^2+3}{(s+1)(s-2)(s-4)}\right\} = \frac{4}{15}e^{-t} - \frac{7}{6}e^{2t} + \frac{19}{10}e^{4t}$$

(35-1)

Example 17
Find the inverse Laplace transform of the expression

$$\frac{s+2}{s^3(s-1)^2}$$

(36)

Solution

Equation (36) is expanded into its partial fractions as follows

$$\frac{s+2}{s^3(s-1)^2} = \frac{A}{s^3} + \frac{B}{s^2} + \frac{C}{s} + \frac{D}{(s-1)^2} + \frac{E}{(s-1)}$$

(36-1)

The cross multiplication of the terms in equation (36-1) gives

$$s+2 = A(s-1)^2 + Bs(s-1)^2 + Cs^2(s-1)^2 + Ds^3 + Es^3(s-1)$$

(36-2)

Equating the coefficients of s^4 in equation (36-2) gives

$$0 = C + E$$

(36-2a)

Equating the coefficients of s^3 in equation (36-2) gives

$$0 = B - 2C + D - E$$

(36-2b)

Equating the coefficients of s^2 in equation (36-2) gives

$$0 = A - 2B + C$$

(36-2c)

Equating the coefficients of s in equation (36-2) gives

$$1 = -2A + B$$

(36-2d)

Equating the constants in equation (36-2) gives

$$A = 2$$

(36-2e)

Hence, (36-2e) and (36-2d) give

$$B = 5$$

(36-2f)

Subtracting (36-2f) and (36-2e) into (36-2c), gives

$$C = 8$$

(36-2g)

36

Subtracting (36-2g) into (36-2a), gives

E = - 8   (36-2h)

Finally, equation (36-2b) gives

D = - B + 2C + E = -5 +16 -8 = 3   (36-2i)

Hence, equation (36-1) becomes

$$\frac{s+2}{s^3(s-1)^2} = \frac{2}{s^3} + \frac{5}{s^2} + \frac{8}{s} + \frac{8}{(s-1)^2} - \frac{8}{(s-1)}$$ (37)

$$L^{-1}\left\{\frac{1}{s^n}\right\} = \frac{t^{n-1}}{(n-1)!}$$

And the inverse Laplace of (37) is

$$L^{-1}\left\{\frac{s+2}{s^3(s-1)^2}\right\} = 2\frac{t^{3-1}}{2!} + 5\frac{t^{2-1}}{1!} + 8 + 8\,t\,e^t - 8e^t$$

$$L^{-1}\left\{\frac{s+2}{s^3(s-1)^2}\right\} = t^2 + 5\,t + 8 + 8\,e^t(t-1)$$ (37-1)

Example 18

Find the inverse Laplace transform of the expression

$$\frac{s^2+1}{(s-1)(s-2)^2}$$ (38)

Solution

We write equation (38) in its partial fractions as follows

$$\frac{s^2+1}{(s-1)(s-2)^2} = \frac{A}{(s-1)} + \frac{B}{(s-2)^2} + \frac{C}{(s-2)}$$ (38-1)

We will use the zeros of the expressions in determining the three constants.

Multiply both sides of (38-1) by (s-1) and let $s \rightarrow 1$

37

$$\lim_{s \to 1} \frac{s^2 + 1}{(s - 2)^2} = A \quad \text{Therefore, A = 2} \tag{38-2}$$

Multiply both sides of (38-1) by $(s-2)^2$ and let $s \to 2$

$$\lim_{s \to 2} \frac{s^2 + 1}{(s - 1)} = B \quad \text{Therefore, B = 5} \tag{38-3}$$

To find C multiply both sides of (38-1) by s and let $s \to \infty$

$$\lim_{s \to \infty} \frac{s^3 + s}{(s - 1)(s - 2)^2} = \lim_{s \to \infty} \frac{As}{(s - 1)} + \lim_{s \to \infty} \frac{Bs}{(s - 2)^2} + \lim_{s \to \infty} \frac{Cs}{(s - 2)} \tag{38-4}$$

Therefore,

$1 = A + 0 + C$

Which gives C = -1

Hence, equation (38-1) becomes

$$\frac{s^2 + 1}{(s - 1)(s - 2)^2} = \frac{2}{(s - 1)} + \frac{5}{(s - 2)^2} - \frac{1}{(s - 2)} \tag{39}$$

And the inverse Laplace transform is

$$L^{-1} \left\{ \frac{s^2 + 1}{(s - 1)(s - 2)^2} \right\} = 2 e^t + 5 e^{2t} - e^{2t} \tag{39-1}$$

Example 19

Find the inverse Laplace transform of the expression

$$\frac{3s^2 + 9s + 16}{(s + 1)(s^2 + 4s + 13)} \tag{40}$$

Solution

We will arrange the terms in the numerator of equation (40) in such manner that yields elementary fractions that could easily inversed by Laplace transform as follows.

First, split the numerator into two terms in order to generate two fractions.

$$\frac{3s^2 + 9s + 16}{(s+1)(s^2 + 4s + 13)} = \frac{(2s^2 + 5s + 3) + (s^2 + 4s + 13)}{(s+1)(s^2 + 4s + 13)}$$

$$= \frac{(s+1)(2s+3) + (s^2 + 4s + 13)}{(s+1)(s^2 + 4s + 13)}$$

$$= \frac{1}{(s+1)} + \frac{(2s+3)}{(s^2 + 4s + 13)}$$

Second, arrange the fraction on the RHS to conform to elementary fractions as follows

$$= \frac{1}{(s+1)} + \frac{(2s+4) - 1}{(s^2 + 4s + 4) + 9}$$

$$= \frac{1}{(s+1)} + \frac{2(s+2)}{(s+2)^2 + 9} - \frac{1}{(s+2)^2 + 9} \qquad (40\text{-}1)$$

Finally, equation (40-1) offers elementary functions that could be easily inversed by Laplace transformation as follows.

$$L^{-1}\left\{\frac{3s^2 + 9s + 16}{(s+1)(s^2 + 4s + 13)}\right\} = L^{-1}\left\{\frac{1}{(s+1)}\right\} + 2L^{-1}\left\{\frac{(s+2)}{(s+2)^2 + 9}\right\} - L^{-1}\left\{\frac{1}{(s+2)^2 + 9}\right\}$$

$$= e^{-t} + 2e^{-2t}\cos 3t - \frac{1}{3}e^{-2t}\sin 3t \qquad (40\text{-}2)$$

To review our steps, in obtaining equation (40-2), we used the following rules
(1) The additive nature of the inverse Laplace transforms
(2) The first shift theorem
(3) The multiplication of the inverse Laplace transform by a constant
(4) The inverse Laplace transforms of a constant, sine, and cosine functions.

Example 20

Find the inverse Laplace transform of the expression

$$\frac{2s+5}{(s^2 + 6s + 25)^2} \qquad (41)$$

Solution

We will arrange equation (41) to yield elementary fractions as follows

$$\frac{2s+5}{(s^2 + 6s + 25)^2} = \frac{2(s+3) - 1}{[(s+3)^2 + 16]^2}$$

39

$$= \frac{2(s+3)}{[(s+3)^2+16]^2} - \frac{1}{[(s+3)^2+16]^2} \tag{41-1}$$

We will use the following inverse Laplace transforms of elementary functions

$$L^{-1}\left\{ \frac{s}{[s^2+k^2]^2} \right\} = \frac{1}{2k} t \sin kt \tag{41-1a}$$

$$L^{-1}\left\{ \frac{1}{[s^2+k^2]^2} \right\} = \frac{1}{2k^3}(\sin kt + k t \cos kt) \tag{41-1b}$$

In addition to the two properties in (41-1a) and (41-1b), we will also use the "first-shift theorem" due to the presence of constants next to "s".

$$L^{-1}\left\{ \frac{2s+5}{(s^2+6s+25)^2} \right\} = L^{-1}\left\{ \frac{2(s+3)}{[(s+3)^2+16]^2} \right\} - L^{-1}\left\{ \frac{1}{[(s+3)^2+16]^2} \right\} \tag{41-2}$$

$$= \frac{2}{8}e^{-3t} t \sin 4t - \frac{1}{128}e^{-3t}(\sin 4t - 4t \cos 4t)$$

$$= \frac{e^{-3t}}{128}(32 t \sin 4t - \sin 4t + 4t \cos 4t)$$

Example 21

Find the inverse Laplace transform of the expression

$$\frac{1}{(s^2+a^2)(s^2+b^2)}, \qquad a^2 \neq b^2 \tag{42}$$

Solution

Equation (42), written in terms of its partial fractions gives

$$\frac{1}{(s^2+a^2)(s^2+b^2)} = \frac{A}{(s^2+a^2)} + \frac{B}{(s^2+b^2)} \tag{42-1}$$

Where the constants A and B are evaluated as before, by equating the coefficients of equal powers of s, as follows.

$1 = A(s^2+b^2) + B(s^2+a^2)$
$0 = A + B$
$1 = Ab^2 + Ba^2$

And
$0 = A b^2 + B b^2$
$1 = B(a^2 - b^2)$

Thus,

$$B = 1/(a^2 - b^2) \tag{42-2}$$
And
$$A = -1/(a^2 - b^2) \tag{42-3}$$

Therefore,

$$\frac{1}{(s^2 + a^2)(s^2 + b^2)} = \frac{1}{a^2 - b^2}\left(\frac{-1}{(s^2 + a^2)} + \frac{1}{(s^2 + b^2)}\right) \tag{43}$$

$$L^{-1}\left\{\frac{1}{(s^2 + a^2)(s^2 + b^2)}\right\} = \frac{1}{a^2 - b^2}\left(\sin bt - \sin at\right)$$

Example 22

Find the inverse Laplace transform of the expression

$$\frac{s}{(s^2 + a^2)(s^2 + b^2)}, \qquad a^2 \neq b^2 \tag{44}$$

Solution

$$L^{-1}\left\{\frac{s}{(s^2 + a^2)(s^2 + b^2)}\right\} = \frac{1}{a^2 - b^2}\left(\cos bt - \cos at\right)$$

Example 23

Find the inverse Laplace transform of the expression

$$\frac{s^2}{(s^2 + a^2)(s^2 + b^2)}, \qquad a^2 \neq b^2 \tag{45}$$

Solution

$$L^{-1}\left\{\frac{s^2}{(s^2 + a^2)(s^2 + b^2)}\right\} = \frac{1}{a^2 - b^2}\left(L^{-1}\left\{\frac{s^2}{(s^2 + a^2)}\right\} - L^{-1}\left\{\frac{s^2}{(s^2 + b^2)}\right\}\right) \tag{45-1}$$

We will arrange the terms in (45-1) to yield elementary terms that could be easily transformed as follows.

$$L^{-1}\left\{\frac{s^2}{(s^2 + a^2)(s^2 + b^2)}\right\} = \frac{1}{a^2 - b^2}\left(L^{-1}\left\{\frac{s^2 + a^2 - a^2}{(s^2 + a^2)}\right\} - L^{-1}\left\{\frac{s^2 + b^2 - b^2}{(s^2 + b^2)}\right\}\right)$$

$$= \frac{1}{a^2 - b^2} \left(L^{-1} \left\{ 1 - \frac{a^2}{(s^2 + a^2)} \right\} - L^{-1} \left\{ 1 - \frac{b^2}{(s^2 + b^2)} \right\} \right)$$

$$= \frac{1}{a^2 - b^2} \left(a\, L^{-1} \left\{ - \frac{a}{(s^2 + a^2)} \right\} - b\, L^{-1} \left\{ - \frac{b}{(s^2 + b^2)} \right\} \right)$$

$$= \frac{1}{a^2 - b^2} \left(a \sin at - b \sin bt \right)$$

1.16. Laplace's solution of linear differential equations with constant coefficients

We will extensively use equation (26), rewritten below for review, throughout this section to solve linear differential equation.

$$L\left\{ F^{(n)}(t) \right\} = s^n\, f(s) - s^{n-1}\, F(0) - s^{n-2}\, F'(0) - \dots F^{(n-1)}(0)$$

Example 24

Solve the equation

$$Y'''(t) - 6\, Y''(t) + 11 Y'(t) - 6 Y(t) = 1 \tag{46}$$

Given that $Y(0) = Y'(0) = Y''(0) = 0$.

Solution

Transforming both sides of equation (46) by Laplace integrals (i.e., multiplying both sides by e^{-st} and integrating w.r.t. t, between 0 and ∞) we get :

$$L\{Y'''(t) - 6\, Y''(t) + 11 Y'(t) - 6 Y(t)\} = L\{1\} \tag{46-1}$$

We will use the property of Laplace transform that equates multiplication by s with integration as follows.

$$s^3\, y(s) - 6\, s^2\, y(s) + 11\, s\, y(s) - 6\, y(s) = 1/s \tag{46-2}$$

Therefore,

$$y(s) = \frac{1}{s\, (s^3 - 6 s^2 + 11 s - 6)} \tag{46-3}$$

We will first determine the roots of the denominator as follows.

$s^3 - 6s^2 + 11s - 6 = (s-1)(s^2 + bs + c)$ (46-4)

Where, the two constants b and c are determined by equating the coefficients of equal powers such that
-6 = b-1
11 = c -b
6 = c
b = -5

Therefore, we get

$s^3 - 6s^2 + 11s - 6 = (s-1)(s^2 - 5s + 6) = (s-1)(s-2)(s-3)$ (46-5)

Equation (46-3) is written as

$$y(s) = \frac{A}{s} + \frac{B}{(s-1)} + \frac{C}{(s-2)} + \frac{D}{(s-3)}$$ (46-6)

Where,
s = 0:
$A(s-1)(s-2)(s-3) = A(0-1)(0-2)(0-3) = -6A = 1$
$A = -1/6$

s = 1:
$Bs(s-2)(s-3) = B(1)(1-2)(1-3) = 2B = 1$
$B = 1/2$

s = 2:
$Cs(s-1)(s-3) = C(2)(2-1)(2-3) = C(-2) = 1$
$C = -1/2$

s = 3:
$Ds(s-1)(s-2) = D(3)(3-1)(3-2) = 6D = 1$
$D = 1/6$

$$y(s) = -\frac{1}{6s} + \frac{1}{2(s-1)} - \frac{1}{2(s-2)} + \frac{1}{6(s-3)}$$ (46-7)

Performing the inverse transformation of (46-7), we get

$$Y(t) = L^{-1}\{y(s)\} = -\frac{1}{6} + \frac{1}{2}e^t - \frac{1}{2}e^{2t} + \frac{1}{6}e^{3t}$$ (46-8)

Example 25

Solve the equation

$Y''(t) — 3\ Y'(t) + 2\ Y(t) = e^t$ (47)

Given that $Y(0) = Y'(0) = 0$

Solution

By using equation (46), the Laplace transform of equation (47), with the given initial conditions, is

$L\ \{Y''(t) — 3\ Y'(t) + 2\ Y(t)\ \} = L\ \{e^t\}$ (47-1)

$(s^2 - 3\ s + 2)\ y(s) = 1/\ (s\text{-}1)$ (47-2)

Therefore,

$$y(s) = \frac{1}{(s\text{-}1)^2(s-2)}$$

$$= \frac{A}{(s\text{-}1)^2} + \frac{B}{(s\text{-}1)} + \frac{C}{(s-2)} = -\frac{1}{(s\text{-}1)^2} - \frac{1}{(s\text{-}1)} + \frac{1}{(s-2)}$$ (47-3)

$$Y(t) = L^{-1}\ \{y(s)\} = -t\,e^t - e^t + e^{2t}$$ (47-4)

Example 26

Solve the equation

$X''(t) + 2\ X'(t) + X(t) = 3\ t\ e^{-t}$ (48)

Given that $X(0) = 4,\ \ X'(0) = 2$

Solution

Using equations (24) and (25) to obtain the Laplace transform of equation (48), we get

$$\left[s^2\ x(s) - s\ X(0) - X'(0)\right] + 2\left[s\ x(s) - X(0)\right] + x(s) = \frac{3}{(s+1)^2}$$ (48-1)

Therefore,

$$s^2\ x(s) - 4\ s\ -2 + 2\ s\ x(s) - 8 + x(s) = \frac{3}{(s+1)^2}$$ (48-2)

$$\left(s^2 + 2\ s + 1\right)\ x(s) = 4\ s + \frac{3}{(s+1)^2} + 10$$ (48-3)

Equation (48-3) gives the Laplace transform as follows

$$x(s) = \frac{4s+10}{\left(s^2+2s+1\right)} + \frac{3}{(s+1)^2\left(s^2+2s+1\right)}$$

Arranging the terms farther we get

$$= \frac{4s+10}{(s+1)^2} + \frac{3}{(s+1)^2(s+1)^2}$$

$$= \frac{4s+4+6}{(s+1)^2} + \frac{3}{(s+1)^2(s+1)^2}$$

$$= \frac{4}{(s+1)} + \frac{6}{(s+1)^2} + \frac{3}{(s+1)^4} \qquad (48\text{-}4)$$

The inverse Laplace transform of equation (48-4), we get

$$L^{-1}\left\{\frac{4}{(s+1)} + \frac{6}{(s+1)^2} + \frac{3}{(s+1)^4}\right\} = 4e^{-t} + 6te^{-t} + 3e^{-t}\frac{t^3}{3!}$$

$$= e^{-t}\left(4 + 6t + \frac{t^3}{2}\right) \qquad (48\text{-}5)$$

$$Y(t) = e^{-t}\left(4 + 6t + \frac{t^3}{2}\right)$$

Example 27

Solve the equation

$$Y''(t) + 6\,Y'(t) + 9\,Y(t) = 6\,t^3\,e^{-3t} \qquad (49)$$

Given that $Y(0) = Y'(0) = 0$

Solution

By using equation (46), the Laplace transform of equation (49), with the given initial conditions, is

$$\left(s^2+6s+9\right)y(s) = \frac{12}{(s+3)^3} \qquad (49\text{-}1)$$

Therefore,

$$y(s) = \frac{12}{(s+3)^5} \tag{49-2}$$

The inverse Laplace transform of (49-2) is

$$Y(t) = L^{-1}\{y(s)\} = 12 e^{-3t} \frac{t^4}{4!} = \frac{1}{2} t^4 e^{-3t} \tag{49-3}$$

Example 28

Solve the equation

$$Y''(t) + 2\,Y'(t) + Y(t) = t \tag{50}$$

Given that $Y(0) = -3$, $Y(1) = -1$

Solution

By using equation (46), the Laplace transform of equation (50), with the given initial conditions, is

$$\left[s^2 \, y(s) - s \, Y(0) - Y'(0) \right] + 2 \left[s \, y(s) - Y(0) \right] + y(s) = \frac{1}{s^2}$$

Put $Y'(0) = B$

$$s^2 \, y(s) + 3s - B + 2s \, y(s) + 6 + y(s) = \frac{1}{s^2}$$

$$\left(s^2 + 2s + 1 \right) y(s) = \frac{1}{s^2} - 3s - 6 + B$$

$$y(s) = \frac{1}{s^2 \left(s^2 + 2s + 1 \right)} + \frac{-3s - 6 + B}{\left(s^2 + 2s + 1 \right)} \tag{50-1}$$

$$= \frac{1}{s^2 \left(s + 1 \right)^2} + \frac{-3s - 3 - 3 + B}{\left(s + 1 \right)^2}$$

$$= \frac{1}{s^2 \left(s + 1 \right)^2} - \frac{3}{\left(s + 1 \right)} + \frac{B - 3}{\left(s + 1 \right)^2} \tag{50-2}$$

We will expand the first fraction on the RHS in terms of its partial fractions as follows

$$\frac{1}{s^2 \left(s + 1 \right)^2} = \frac{A}{s^2} + \frac{C}{s} + \frac{D}{\left(s + 1 \right)^2} + \frac{E}{\left(s + 1 \right)} \tag{50-3}$$

The constants in (50-3) are determined as follows.

46

Cross multiplication of the terms in (50-3):

$$1 = A(s+1)^2 + Cs(s+1)^2 + D s^2 + E s^2 (s+1)$$

Coefficients of s^3 $: 0 = C+E$
Coefficients of s^2 $: 0 = A+2C+D+E$
Coefficients of s $: 0 = 2A+C$
Coefficients of constants: $1 = A$

Therefore,

$C = -2$
$E = 2$
$D = 1$
Thus, equation (50-3) becomes

$$\frac{1}{s^2(s+1)^2} = \frac{1}{s^2} - \frac{2}{s} + \frac{1}{(s+1)^2} + \frac{2}{(s+1)} \tag{50-4}$$

Substituting equation (50-4) in (50-2), we get

$$y(s) = -\frac{3}{(s+1)} + \frac{B-3}{(s+1)^2} + \frac{1}{s^2} - \frac{2}{s} + \frac{1}{(s+1)^2} + \frac{2}{(s+1)}$$

$$= \frac{1}{s^2} - \frac{2}{s} + \frac{B-2}{(s+1)^2} - \frac{1}{(s+1)} \tag{50-5}$$

The inverse Laplace transform of (50-5) is

$$Y(t) = L^{-1}\left\{ \frac{1}{s^2} - \frac{2}{s} + \frac{B-2}{(s+1)^2} - \frac{1}{(s+1)} \right\} = t - 2 + (B-2) t e^{-t} - e^{-t} \tag{50-6}$$

Since $Y(1) = -1$, then (50-6) gives

$-1 = 1 - 2 + (B-2) e^{-1} - e^{-1}$
Therefore,

$B = 3$

Substituting by $B = 3$ in (50-6), we get

$$Y(t) = t - 2 + (t-1) e^{-t} \tag{50-7}$$

Example 29

Solve the equation

$$Y''' (t) + Y(t) = 1 \tag{51}$$

47

Given that $Y(0) = Y'(0) = Y''(0) = 0$

Solution

By using equation (46), the Laplace transform of equation (51), with the given initial conditions, is

$$(s^3 + 1) y(s) = \frac{1}{s}$$

Therefore,

$$y(s) = \frac{1}{s(s^3 + 1)} \qquad (51\text{-}1)$$

In order to transform the cubic polynomial in (51-1) via Laplace integration, we first need to represent the cubic polynomial into its partial multiplicative polynomial or quadratic and linear orders as follows. (We assume that the reader prefers direct proof rather than resorting to **corollaries** of the binomial theory).

$$s^3 + 1 = (s + 1)(s^2 + As + B) \qquad (51\text{-}2)$$

The two constants A and B are obtained by equating the coefficients of terms of equal powers. Thus

Equating the coefficients of s gives: $0 = A+1$
Equating the constants gives: $1 = B$

Therefore, equation (51-2) becomes

$$s^3 + 1 = (s + 1)(s^2 - s + 1) \qquad (51\text{-}2a)$$

Farther arrangements of the rightmost term in (51-2a) renders it conforming to the elementary Laplace functions as follows

$$s^3 + 1 = (s + 1) [s^2 - s + (1/4) + (3/4)]$$
$$= (s + 1) [\{s-(1/2)\}^2 + (3/4)] \qquad (51\text{-}2b)$$

Therefore, by substituting with equation (51-2b) into equation (51-1), the latter is now brought to the user-friendly style of Laplace transformations as follows

$$y(s) = \frac{1}{s(s+1)\left[\left(s - \frac{1}{2}\right)^2 + \frac{3}{4}\right]} \qquad (51\text{-}3)$$

$$y(s) = \frac{A}{s} + \frac{B}{(s+1)} + \frac{Cs+D}{\left(\left(s - \frac{1}{2}\right)^2 + \frac{3}{4}\right)}$$

(51-3a)

Determining the constants in equation (51-3a) from the zeros of the polynomials

At s = 0:
A(s+1)([{s-(1/2)}^{2} + (3/4)] =1
A(1)[(1/4) + (3/4)] = 1
A = 1

(51-3b)

At s = -1:
B(s)([{s-(1/2)}^{2} + (3/4)] =1
B(-1)[{-1-(1/2)} ^{2} + (3/4)] = 1
B(-1)[(9/4) +(3/4)] = B(-1)(3) = 1
B = -1/3

(51-3c)

Determining C and D
(1)(s + 1) (s^2 -s + 1) +(-1/3)(s) (s^2 -s + 1) + (Cs + D)(s)(s + 1) = 1
(s + 1) (s^2 -s + 1) +(-1/3)(s) (s^2 -s + 1) + (Cs + D)(s)(s + 1) = 1

Coefficients of s^3:

1 – (1/3) + C = 0
C = -2/3

(51-3d)

Coefficients of s^2:

-1 + 1+ (1/3) + C + D=0
D = 1/3

(51-3e)

Substituting the constants from equations (51-3b), (51-3c), (51-3d), and (51-3e) in equation (51-3a) we get

$$y(s) = \frac{1}{s} - \frac{1}{3(s+1)} + \frac{1}{3}\frac{-2s+1}{\left(\left(s - \frac{1}{2}\right)^2 + \frac{3}{4}\right)}$$

$$= \frac{1}{s} - \frac{1}{3(s+1)} - \frac{2}{3}\frac{s - \frac{1}{2}}{\left(\left(s - \frac{1}{2}\right)^2 + \frac{3}{4}\right)}$$

(51-4)

The inverse Laplace transform of (51-4) is now straightforward via the use of elementary functions as follows.

$$Y(t) = L^{-1}\left\{ \frac{1}{s} - \frac{1}{3(s+1)} - \frac{2}{3}\frac{s-\frac{1}{2}}{\left(\left(s-\frac{1}{2}\right)^2 + \frac{3}{4}\right)} \right\}$$

$$Y(t) = 1 - \frac{1}{3}e^{-t} - \frac{2}{3}e^{\frac{1}{2}t}\cos\sqrt{\frac{3}{4}}\,t \tag{51-5}$$

Example 30

Solve the equation

$$Y''(t) + k^2\,Y(t) = 0 \tag{52}$$

Solution

Assume the initial conditions as follows
Y(0) = A
Y'(0) = B

By using equation (46), the Laplace transform of equation (52), with the assumed initial conditions, is

$$\left[s^2\,y(s) - s\,Y(0) - Y'(0) \right] + k^2\,y(s) = 0 \tag{52-1}$$

$$\left(s^2 + k^2 \right) y(s) = As + B$$

$$y(s) = \frac{As}{s^2 + k^2} + \frac{B}{s^2 + k^2} \tag{52-3}$$

$$Y(t) = L^{-1}\left\{ y(s) \right\} = A\cos kt + \frac{B}{k}\sin kt \tag{52-4}$$

This is an example on **simple harmonic oscillation**, A and B being respectively the initial displacement and velocity.

Example 31

Solve the equation

$$X''(t) + 4\,X(t) = 10\sin 3t \tag{53}$$

Given that X(0) = X'(0) = 0

Solution

By using equation (46), the Laplace transform of equation (53), with the given initial conditions, is

$$\left(s^2 + 4\right) x(s) = 10\left(\frac{3}{s^2 + 9}\right)$$

$$x(s) = \frac{30}{\left(s^2 + 9\right)\left(s^2 + 4\right)}$$

$$= \frac{A}{\left(s^2 + 9\right)} + \frac{B}{\left(s^2 + 4\right)} \qquad (53\text{-}1)$$

Determining the constants A and B

$30 = A\ (s^2 + 9) + B\ (s^2 + 4)$

$0 = A - B$

$30 = 9\ A + 4\ B$

$\quad = 5\ A$

$A = 6$

$B = -6$

Thus, equation (53-1) becomes

$$x(s) = \frac{6}{\left(s^2 + 9\right)} - \frac{6}{\left(s^2 + 4\right)} \qquad (53\text{-}2)$$

$$X(t) = L^{-1}\left\{ x(s) \right\} = 3 \sin 2t - 2 \sin 3t \qquad (53\text{-}3)$$

This is an example on **forced oscillation without damping**.

Example 32

Solve the equation

$Y''(t) + n^2\ Y(t) = a \sin nt$

Given that $Y(0) = Y'(0) = 0$ \qquad (54)

Solution

By using equation (46), the Laplace transform of equation (54), with the given initial conditions, is

$$\left(s^2 + n^2\right) y(s) = \frac{a\ n}{s^2 + n^2}$$

Therefore,

$$y(s) = \frac{a \, n}{\left(s^2 + n^2\right)^2} \qquad (54\text{-}1)$$

$$Y(t) = L^{-1}\{y(s)\} = \frac{1}{2\,k^3}\left(\sin kt - kt \cos kt\right) \qquad (54\text{-}2)$$

$$= \frac{a}{2\,n^2}\left(\sin kt - kt \cos kt\right)$$

This is an example on **resonance** when **the frequency of the impressed oscillations is equal to the natural frequency**.

Example 33

Solve the equation

$$Y''(t) + 4Y'(t) + 13Y(t) = 5 \cos 3t \qquad (55)$$

Given that $Y(0) = 1/4, \qquad Y'(0) = 2$

Solution

By using equation (46), the Laplace transform of equation (55), with the given initial conditions, is

$$\left[s^2\, y(s) - s\, Y(0) - Y'(0)\right] + 4\left[s\, y(s) - Y(0)\right] + 13 y(s) = \frac{5\,s}{s^2 + 9} \qquad (55\text{-}1)$$

Substituting by the initial conditions in (55-1), we get

$$s^2\, y(s) - \frac{1}{4}s - 2 + 4\left[s\, y(s) - \frac{1}{4}\right] + 13\, y(s) = \frac{5\,s}{s^2 + 9}$$

$$(s^2 + 4s + 13)\, y(s) - \frac{1}{4}s - 3 = \frac{5\,s}{s^2 + 9}$$

$$y(s) = \frac{s + 12}{4\,(s^2 + 4s + 13)} + \frac{5\,s}{(s^2 + 9)(s^2 + 4s + 13)} \qquad (55\text{-}2)$$

$$= \frac{(s + 12)(s^2 + 9) + 20\,s}{4\,(s^2 + 4s + 13)(s^2 + 9)} = \frac{s^3 + 12s^2 + 29\,s + 108}{4\,(s^2 + 4s + 13)(s^2 + 9)}$$

$$= \frac{As + B}{4\,(s^2 + 9)} + \frac{Cs + D}{(s^2 + 4s + 13)} \qquad (55\text{-}3)$$

Determining the coefficients in (55-3)

$$(As + B)(s^2 + 4s + 13) + 4\,(Cs + D)(s^2 + 9) = s^3 + 12\,s^2 + 29\,s + 108 \qquad (55\text{-}3a)$$

52

The coefficients of s^3:
(i) A + 4C = 1

C = (1 - A) / 4

The coefficients of s^2:
(ii) 4 A +B + 4D = 12

The coefficients of s:
(iii) 13A + 4B + 36C = 29
 From (i),
13 A + 4B + 36(1-A)/4 = 4A + 4B + 9 = 29

B = 5 -A

The constants:
(iv) 13B + 36D = 108

From (iii)
D = (108-13B)/36 = 3 − 13(5-A)/36 = (43 + 13A)/36

From (ii)

4A + (5-A) + (43+13A)/9 = 12
(43+ 40A) = 63
40A = 20

A = 1/2
B = 5 − 1/2 = 9/2
C = (1-1/2)/4 = 1/8
D = (43 + 13 /2)/36 = 99/72 = 11/8

Substituting with the constants A, B, C, and D in equation (55-3) gives

$$y(s) = \frac{s+9}{8(s^2+9)} + \frac{s+11}{8(s^2+4s+13)} \tag{55-4a}$$

$$= \frac{s}{8(s^2+9)} + \frac{9}{8(s^2+9)} + \frac{s+2}{8(s^2+4s+13)} + \frac{9}{8(s^2+4s+13)}$$

$$= \frac{s}{8(s^2+9)} + \frac{9}{8(s^2+9)} + \frac{s+2}{8[(s+2)^2+9]} + \frac{9}{8[(s+2)^2+9]} \tag{55-4b}$$

Equation (55-4b) represents elementary functions with known inverse Laplace transforms as follows.

$$Y(t) \ = L^{-1}\left\{ \frac{s}{8(s^2+9)} + \frac{9}{8(s^2+9)} + \frac{s+2}{8\,[(s+2)^2+9]} + \frac{9}{8\,[(s+2)^2+9]} \right\}$$

$$= \frac{1}{8}\cos 3t + \frac{3}{8}\sin 3t + \frac{1}{8}e^{-2t}\cos 3t + \frac{3}{8}e^{-2t}\sin 3t$$

$$= \frac{1}{8}(1+e^{-2t})(\cos 3t + 3\sin 3t) \qquad (55\text{-}5)$$

This is an example on **forced oscillations with damping**.

Example 34

Solve the simultaneous differential equations

$$\dot{x} - x + 2y = 0$$

$$\dot{y} - 5x - 3y = 0 \qquad (56)$$

Given that $x(0) = y(0) = 1$

Solution

The Laplace transforms of equation (56) are

$$s\,x(s) - x(0) - x(s) + 2\,y(s) = 0$$
$$s\,y(s) - y(0) - 5\,x(s) - 3\,y(s) = 0 \qquad (56\text{-}1)$$

Substituting in (56-1) by the initial conditions $x(0) = y(0) = 1$, we get

$$(s-1)\,x(s) + 2\,y(s) = 1$$
$$-5\,x(s) + (s-3)\,y(s) = 1 \qquad (56\text{-}2)$$

Eliminating $y(s)$ between equations (56-2), we get

$$x(s) = \frac{s-5}{s^2-4s+13}$$

$$= \frac{(s-2)-3}{(s-2)^2+9}$$

$$= \frac{(s-2)}{(s-2)^2+9} - \frac{3}{(s-2)^2+9} \qquad (56\text{-}3)$$

The inverse Laplace transform of (56-3) is

$$x(t) = L^{-1}\left\{\frac{(s-2)}{(s-2)^2+9} - \frac{3}{(s-2)^2+9}\right\}$$

$$= e^{2t}\cos 3t - \frac{3}{3}e^{2t}\sin 3t$$

$$= e^{2t}(\cos 3t - \sin 3t) \tag{56-4}$$

Eliminating $x(s)$ between equations (56-2), we get

$$5(s-1)x(s) + 10y(s) = 5$$
$$-5(s-1)x(s) + (s-1)(s-3)y(s) = (s-1)$$

Adding the two equations gives

$$[10 + (s-1)(s-3)]y(s) = 5 + (s-1)$$

$$y(s) = \frac{s+4}{[10 + (s-1)(s-3)]}$$

$$= \frac{s+4}{s^2 - 4s + 13} = \frac{s-2+6}{(s-2)^2+9}$$

$$= \frac{s-2}{(s-2)^2+9} + \frac{6}{(s-2)^2+9} \tag{56-5}$$

The inverse Laplace transform of (56-5) is

$$y(t) = L^{-1}\left\{\frac{s-2}{(s-2)^2+9} + \frac{6}{(s-2)^2+9}\right\}$$

$$= e^{2t}(\cos 3t + \sin 3t) \tag{56-6}$$

1.16.1. Summary of the solution of differential equations

(1) An example on competing growth rates

$$Y'''(t) - 6Y''(t) + 11Y'(t) - 6Y(t) = 1 \tag{46}$$

Given that $Y(0) = Y'(0) = Y''(0) = 0$.

$$Y(t) = L^{-1}\{y(s)\} = -\frac{1}{6} + \frac{1}{2}e^t - \frac{1}{2}e^{2t} + \frac{1}{6}e^{3t} \tag{46-8}$$

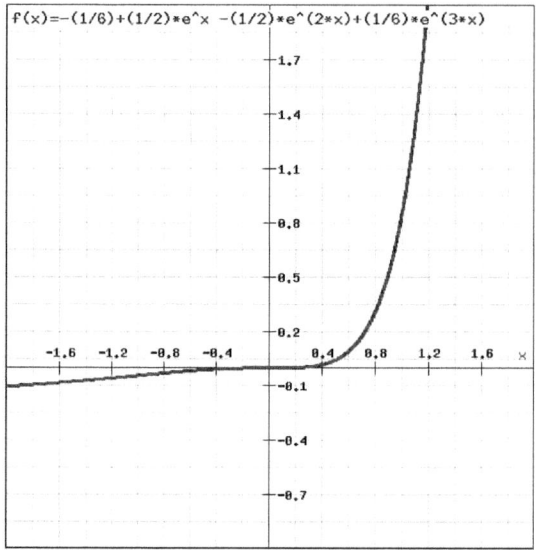

f(x)=-(1/6)+(1/2)*e^x -(1/2)*e^(2*x)+(1/6)*e^(3*x)

(2) An example competing scaled growth rates

$$Y''(t) - 3\,Y'(t) + 2\,Y(t) = e^t \tag{47}$$

Given that $Y(0) = Y'(0) = 0$

$$Y(t) = L^{-1}\{y(s)\} = -t\,e^t - e^t + e^{2t} \tag{47-4}$$

56

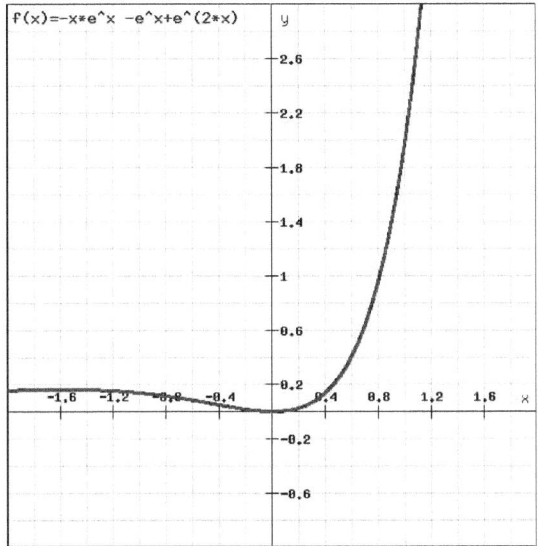

f(x)=-x*e^x -e^x+e^(2*x)

(3) An example on competing decay and growth rates

$$X''(t) + 2\,X'(t) + X(t) = 3\,t\,e^{-t} \tag{48}$$

Given that $X(0) = 4, \quad X'(0) = 2$

$$Y(t) = e^{-t}\left(4 + 6\,t + \frac{t^3}{2}\right)$$

57

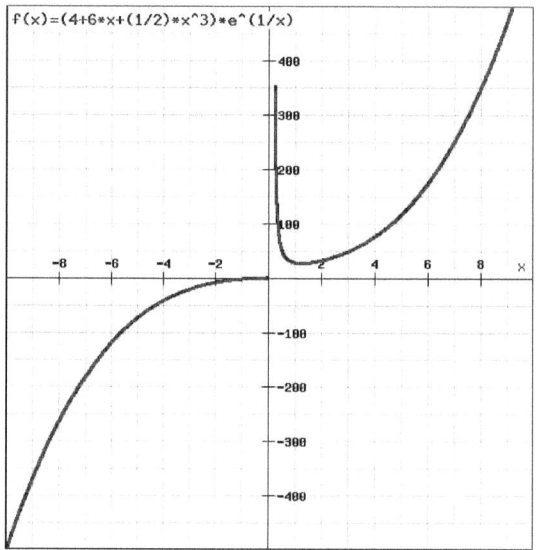

f(x)=(4+6*x+(1/2)*x^3)*e^(1/x)

(4) An example on competing acute decay and growth rates

$$Y''(t) + 6\,Y'(t) + 9\,Y(t) = 6\,t^3\,e^{-3t} \tag{49}$$

Given that $Y(0) = Y'(0) = 0$

$$Y(t) = L^{-1}\{y(s)\} = 12\,e^{-3t}\,\frac{t^4}{4!} = \frac{1}{2}\,t^4 e^{-3t} \tag{49-3}$$

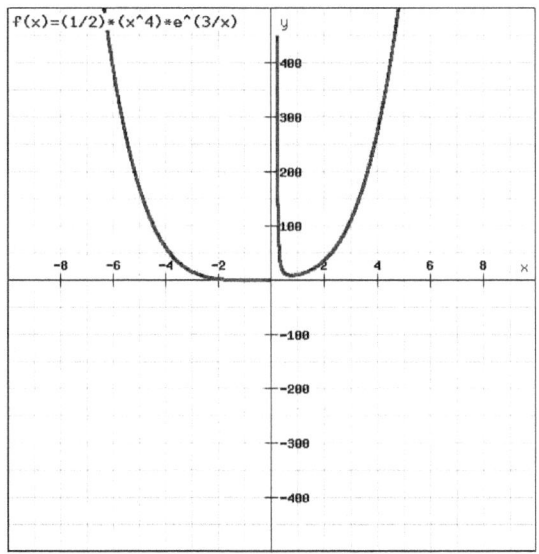

(5) An example on asymptotic growth rate

$$Y''(t) + 2\,Y'(t) + Y(t) = t \qquad\qquad (50)$$

Given that $Y(0) = -3$, $Y(1) = -1$
$$Y(t) = t - 2 + (t-1)\,e^{-t} \qquad\qquad (50\text{-}7)$$

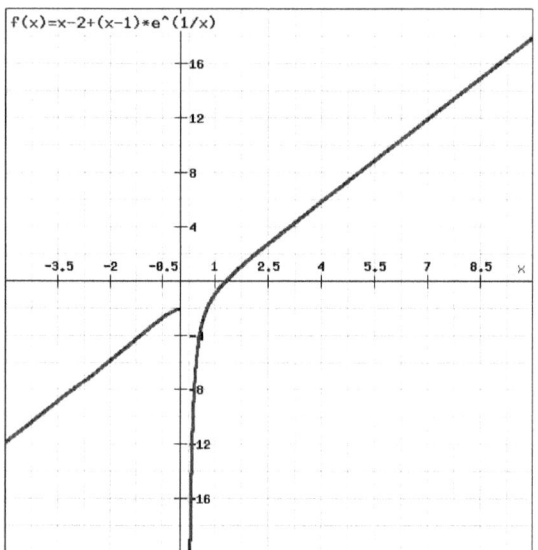

f(x)=x-2+(x-1)*e^(1/x)

(6) Combined growth, decay, and oscillation

$$Y'''(t) + Y(t) = 1 \tag{51}$$

Given that $Y(0) = Y'(0) = Y''(0) = 0$

$$Y(t) = 1 - \frac{1}{3}e^{-t} - \frac{2}{3}e^{\frac{1}{2}t}\cos\sqrt{\frac{3}{4}}t \tag{51-5}$$

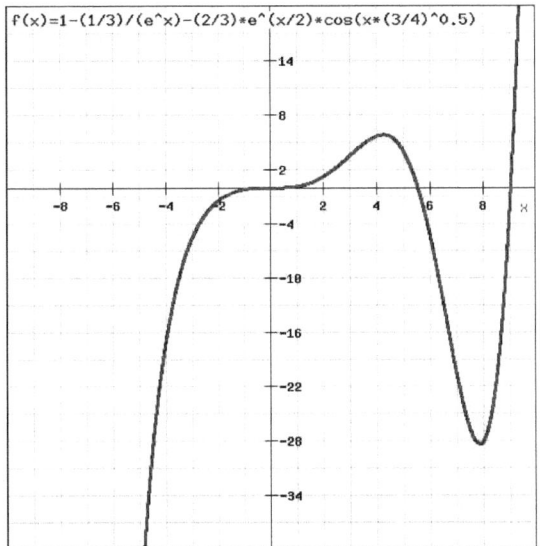

f(x)=1-(1/3)/(e^x)-(2/3)*e^(x/2)*cos(x*(3/4)^0.5)

(7) An example on **simple harmonic oscillation**

Y"(t) + k² Y(t) = 0

$$Y(t) = L^{-1}\{y(s)\} = A\cos kt + \frac{B}{k}\sin kt \qquad (52\text{-}4)$$

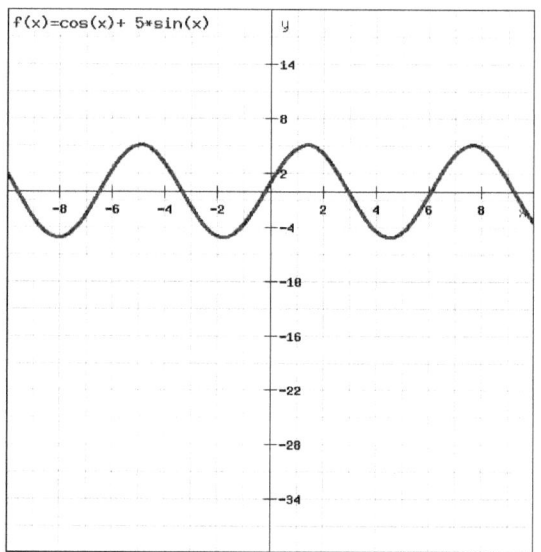

f(x)=cos(x)+ 5*sin(x)

(8) An example on **forced oscillation without damping**

X"(t) + 4 X(t) = 10 sin 3t   (53)

Given that X(0) = X'(0) = 0

$$X(t) = L^{-1} \{ x(s) \} = 3 \sin 2t - 2 \sin 3t \qquad (53\text{-}3)$$

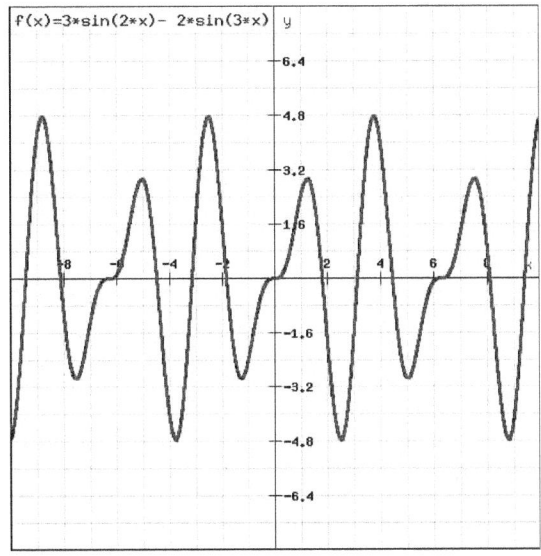

f(x)=3*sin(2*x)- 2*sin(3*x)

(9) An example on **resonance** when **the frequency of the impressed oscillations is equal to the natural frequency.**

$Y''(t) + n^2 Y(t) = a \sin nt$

Given that $Y(0) = Y'(0) = 0$

$$Y(t) = L^{-1}\{ y(s) \} = \frac{1}{2 k^3} (\sin kt - kt \cos kt)$$ (54-2)

$$= \frac{a}{2 n^2} (\sin kt - kt \cos kt)$$

63

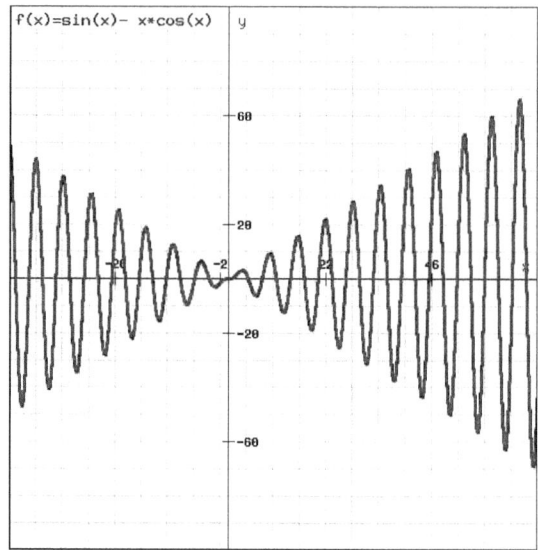

(10) An example on **forced oscillations with damping**

$Y''(t) + 4Y'(t) + 13Y(t) = 5 \cos 3t$ (55)

Given that $Y(0) = 1 / 4, \qquad Y'(0) = 2$

$$Y(t) = L^{-1}\left\{\frac{s}{8(s^2+9)} + \frac{9}{8(s^2+9)} + \frac{s+2}{8[(s+2)^2+9]} + \frac{9}{8[(s+2)^2+9]}\right\}$$

$$= \frac{1}{8}\cos 3t + \frac{3}{8}\sin 3t + \frac{1}{8}e^{-2t}\cos 3t + \frac{3}{8}e^{-2t}\sin 3t$$

$$= \frac{1}{8}(1 + e^{-2t})(\cos 3t + 3\sin 3t)$$ (55-5)

64

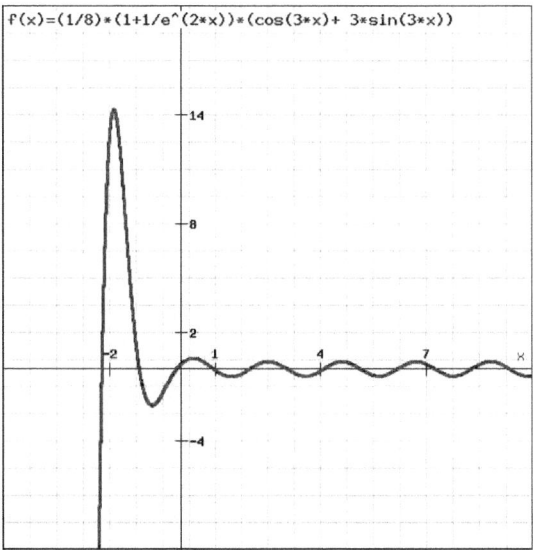

f(x)=(1/8)*(1+1/e^(2*x))*(cos(3*x)+ 3*sin(3*x))

(11) Initial oscillation followed by exponential growth

$$\dot{x} - x + 2y = 0$$

$$\dot{y} - 5x - 3y = 0 \tag{56}$$

Given that $x(0) = y(0) = 1$

$$x(t) = e^{2t} (\cos 3t - \sin 3t) \tag{56-4}$$

$$y(t) = e^{2t} (\cos 3t + \sin 3t) \tag{56-6}$$

65

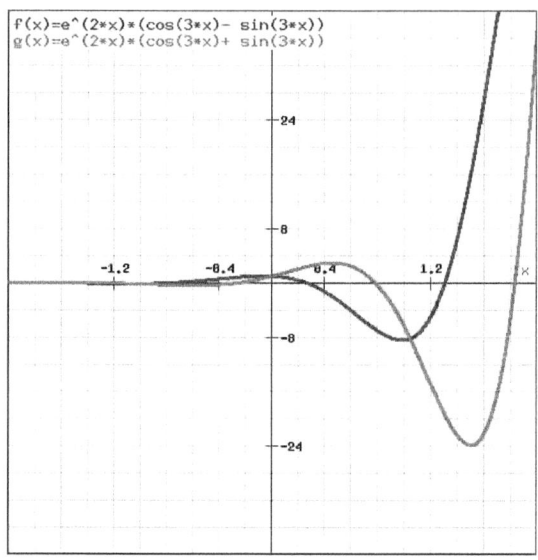

f(x)=e^(2*x)*(cos(3*x)- sin(3*x))
g(x)=e^(2*x)*(cos(3*x)+ sin(3*x))

1.17. Exercises on Laplace transformation

Find the Laplace transform of each of the following functions

(1) $3 e^{5t}$

(2) $4 e^{-3t}$

(3) $4t - 5$

(4) $2 \cos 6t$

(5) $6 \sin 2t - 5 \cos 2t$

(6) $t^2 - 3t + 5$

(7) $\cos^2 kt$

(8) $(\sin t - \cos t)^2$

(9) $\cosh^2 4t$

(10) $\sin t + 3 \cos t$

(11) $F(t) = 4$ in the interval $0 < t < 1$
         $= 3$ $t > 1$

(12) $\Phi(t) = \sin 2t$ in the interval $0 < t < \pi$
          $= 0$ $t > \pi$

(13) $t^{5/2}$

(14) $t^3 e^{-2t}$

(15) $e^{-t} \cos 4t$

(16) $e^{2t} \sin t$

(17) $e^{-4t} \cosh 2t$

Find the inverse Laplace transform of the following functions

(18) $\dfrac{15}{s^2 + 4s + 13}$

(19) $\dfrac{s+1}{s^2+6s+25}$

(20) $\dfrac{s}{s^2-8s+16}$

(21) $\dfrac{1}{s^2+2s+5}$

(22) $\dfrac{2}{s^2-6s+13}$

(23) $\dfrac{1}{s^2+8s+16}$

(24) $\dfrac{s-5}{s^2+6s+13}$

(25) $\dfrac{3s+1}{(s+1)^4}$

(26) $\dfrac{s^2}{(s+2)^3}$

(27) $\dfrac{s+2}{s^2-6s+8}$

(28) $\dfrac{2s^2+5s-4}{s^3+s^2-2s}$

(29) $\dfrac{2s^2+1}{s(s^2+1)}$

(30) $\dfrac{1}{s^3(s^2+1)}$

(31) $\dfrac{5s-2}{s^2(s+2)(s-1)}$

(32) $\dfrac{s^2+s-4}{(s^2-2s+2)(s^2+2s-3)}$

(33) $\dfrac{s}{(s^2+1)(s^2+3)}$

(34) $\dfrac{5s+3}{(s-1)(s^2+2s+5)}$

(35) $\dfrac{s+1}{(s^2+2s+2)^2}$

(36) $\dfrac{s+1}{s(s^2+s-6)}$

Solve the following differential equations

(37) $Y'(t) - Y(t) = 4\,e^t\cos t$, given that $Y(0) = 3$

(38) $Y''(t) - 5Y'(t) + 6\,Y(t) = e^{2t}$, given that $Y(0) = 1$, $Y'(0) = 0$

(39) $Y''(t) + 2Y'(t) + 5\,Y(t) = e^{-t}\sin t$, given that $Y(0) = 0$

(40) $Y^{(4)}(t) + 4Y^{(3)}(t) + 4\,Y''(t) = 0$, given that $Y(0) = 1$ and $Y'(0) = Y''(0) = Y'''(0) = 0$

(41) $Y''(t) + Y(t) = t$, given that $Y(0) = 1$ and $Y'(0) = -2$

(42) $Y''(t) - 3Y'(t) + 2\,Y(t) = 4\,e^{2t}$, given that $Y(0) = -3$

(43) $Y''(t) + 9Y(t) = \cos 2t$, given that $Y(0) = 1$, $Y(\dfrac{\pi}{2}) = -1$

(44) $Y''(t) + 2Y'(t) + Y(t) = \sin 2t$, given that $Y(0) = Y'(0) = 0$

(45) $Y'(t) + Y(t) = t^2\,e^{-t}$, given that $Y(0) = Y_o$

(46) $x''(t) + 4\,x'(t) + 4\,x(t) = 4\,e^{-2t}$, given that $x(0) = -1$

(47) $x''(t) + x(t) = 6\cos 2t$, given that $x(0) = 3$, $x'(0) = 1$

(48) $Y'''(t) - 3\,Y''(t) + 3\,Y'(t) - Y(t) = t^2\,e^t$, given that $Y(0) = 1$, $Y'(0) = 0$, $Y''(0) = -2$

(49) $y''(x) + 9\,y(x) = 40\,e^x$, given that $y(0) = 5$, $y'(0) = -2$

(50) $x''(t) - 4\,x'(t) + 4\,x(t) = e^{2t}$, given that $x'(0) = 0$, $x(1) = 0$

(51) $x'''(t) - 3\,x''(t) + 3\,x'(t) - x(t) = 16\,e^{3t}$, given that $x(0) = 0$, $x'(0) = 4$, $x''(0) = 6$

69

Solve the following simultaneous equations

(52) 3 X' (t) + 2 X (t) - Y(t) = t
    2 Y' (t) - X (t) + Y(t) = 5 e⁻ᵗ
    Given that X(0) = Y(0) = 0

(53) (D-2) x + 3 y = 0
    2 x + (D-1) y = 0
    Given that x(0) = 8, y(0) = 3

(54) x" (t) - x (t) + 5 y' = t
    y" (t) – 4 y(t) -2 x'(t) = -2
    Given that x(0) = x'(0) = y(0) = y'(0) = 0

(54) $5\ddot{x} - 2\ddot{y}$ (t) + 4 x - y = e⁻ᵗ

    $\ddot{x} + 8 x - 3 y = 5 e^{-t}$
    Given that x = y = 0 where t = 0

Answers

(1) $\dfrac{3}{s-5}$, s > 5

(2) $\dfrac{4}{s+3}$, s > -3

(3) $\dfrac{4}{s^2} - \dfrac{5}{s}$, s > 0

(4) $\dfrac{2s}{s^2+36}$, s > 0

(5) $\dfrac{12-5s}{s^2+4}$, s > 0

(6) $\dfrac{2}{s^3} - \dfrac{3}{s^2} + \dfrac{5}{s}$, s > 0

(7) $\dfrac{s^2+2k^2}{s(s^2+4k^2)}$, s > 0

(8) $\dfrac{s^2-2s+4}{s(s^2+4)}$, s > 0

(9) $\dfrac{s^2-32}{s(s^2-64)}$, s > | 8 |

(10) $\dfrac{1+3s}{s^2+1}$, s > 0

(11) $\dfrac{1}{s}(4 - e^{-s})$, $s > 0$

(12) $\dfrac{2(1 - e^{-\pi s})}{s^2 + 4}$, $s > 0$

(13) $\dfrac{15}{8 s^3}\left(\dfrac{\pi}{s}\right)^{\frac{1}{2}}$, $s > 0$

(14) $\dfrac{6}{(s + 2)^4}$

(15) $\dfrac{s + 1}{s^2 + 2s + 17}$

(16) $\dfrac{1}{s^2 - 4s + 5}$

(17) $\dfrac{s + 4}{s^2 + 8s + 12}$

(18) $5 e^{-2t} \sin 3t$

(19) $e^{-3t}(\cos 4t + \dfrac{1}{4} \sin 4t)$

(20) $e^{4t}(1 + 4t)$

(21) $\dfrac{1}{2} e^{-t} \sin 2t$

(22) $e^{3t}(\cos 2t + \dfrac{3}{2} \sin 2t)$

(23) $t e^{-4t}$

(24) $e^{-3t}(\cos 2t - 4 \sin 2t)$

(25) $e^{-t}(\dfrac{3}{2} t^2 - \dfrac{1}{3} t^3)$

(26) $e^{-2t}(1 - 4t + 2t^2)$

(27) $3 e^{4t} - 2 e^{2t}$

(28) $2 + e^t - e^{-2t}$

(29) $1 + e^{-t} - 3 t e^{-t}$

(30) $\dfrac{1}{2} t^2 - 1 + \cos t$

(31) $t - 2 + e^t + e^{2t}$

71

(32) $e^t (\cos t + \sin t) - e^{-t} \sinh 2t$

(33) $\dfrac{1}{2} (\cos t - \cos \sqrt{3}\, t)$

(34) $e^t - e^{-t}(\cos 2t - \dfrac{3}{2} \sin 2t)$

(35) $\dfrac{1}{2} t\, e^{-t} \sin t$

(36) $-\dfrac{1}{6} + \dfrac{3}{10}\, e^{2t} - \dfrac{2}{15}\, e^{-3t}$

(37) $Y(t) = 3\, e^t + 4\, e^t \sin t$

(38) $Y(t) = (2-t)\, e^{2t} - e^{3t}$

(39) $Y(t) = (1/3)\, e^{-t}(\sin t + \sin 2t)$

(40) $Y(t) = 1$

(41) $Y(t) = t + \cos t - 3 \sin t$

(42) $Y(t) = -7\, e^t + 4\, e^{4t} + + 4\, e^{2t}$

(43) $Y(t) = (4/5) \cos 3t + (4/5) \sin 3t + (1/5) \cos 2t$

(44) $Y(t) = (4/25)\, e^{-t} + (2/5)\, t\, e^{-t} - (4/25) \cos 2t - (3/25) \sin 2t$

(45) $Y(t) = (1/3)\, t^3\, e^{-t} + Y_o\, e^{-t}$

(46) $Y(t) = (2\, t^2 + 2\, t - 1)\, e^{-2t}$

(47) $x(t) = 5 \cos t + \sin t - 2 \cos 2t$

(48) $Y(t) = [\, 1 - t - (1/2)\, t^2 + (1/60)\, t^5\,]\, e^t$

(49) $x(t) = 4\, e^t + \cos 3x - 2 \sin 3x$

(50) $x(t) = (1/2)\, (1-t)^2\, e^{2t}$

(51) $x(t) = 2\, e^{3t} - (\, 5t^2 + 2\,)\, e^t$

(52) $X(t) = 6\, e^{-t/6} - (t+1)\, e^{-t} + t - 5$
$\quad\quad Y(t) = 9\, e^{-t/6} + (t-2)\, e^{-t} + t - 7$

(53) $x = 5\,e^{-t} + 3\,e^{4t}$
$\quad y = 5\,e^{-t} - 2\,e^{4t}$

(54) $x(t) = -t + 5\sin t - 2\sin 2t$
$\quad y(t) = 1 - 2\cos t + \cos 2t$

(55) $x = 2\,e^{-t} + e^{t} - 3\,e^{-2t}$
$\quad y = 3\,e^{-t} + 3\,e^{t} - 6\,e^{-2t}$

CHAPTER 2

GENERAL THEOREMS ON THE LAPLACE TRANSFORMATION

2.1. The unit step function

This is defined as a function that vanishes when t < **a**, and becomes unity when t > **a**, where **a** being a positive constant.

Let this function be denoted by U(t—**a**), then

$$U(t-a) \quad = 0 \quad t < a \qquad\qquad (57)$$
$$= 1 \quad t > a$$

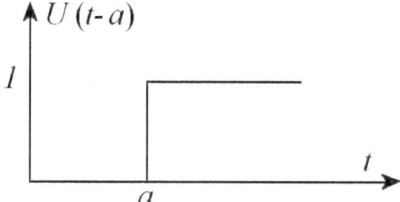

Let us obtain the Laplace transform of this function,

$$L\left\{U(t-0)\right\} = \int_0^a e^{-st}\, 0 \, dt + \int_0^a e^{-st}\, 0 \, dt = \frac{e^{-as}}{a}, \quad s > 0 \qquad (57\text{-}1)$$

If **a** = 0 we have the following special case

$$L\left\{U(t)\right\} = 1/s,$$

Where, U(t) = 0, t < 0 and U(t) = 1, t > 0

Any function F(t) multiplied by U(t—a) will have a value zero for t < a and F(t) for t > a.

i.e.,

$$U(t-a)\,F(t) \quad = 0 \quad t < a \qquad\qquad (57\text{-}2)$$
$$= F(t) \quad t > a$$

2.2. The second translation or shifting property

We shall now show that, if **a** is a positive constant and if

$$L^{-1} \{ f(s) \} = F(t)$$

Then,

$$L^{-1} \{ e^{-as} f(s) \} = 0, \qquad 0 < t < a$$
$$= F(t-a), \qquad t > a$$

This can also be written according to the in the form

$$L^{-1} \{ e^{-st} f(s) \} = U(t-a) F(t-a), \qquad s > 0 \tag{57-3}$$

For

$$e^{-as} f(s) = e^{-as} \int_0^\infty e^{-st} F(t) \, dt = \int_0^\infty e^{-s(t+a)} F(t) \, dt \tag{57-4}$$

Put

$$t + a = \tau,$$

Then

$$dt = d\tau$$

Therefore,

$$\int_0^\infty e^{-s(t+a)} F(t) \, dt = \int_0^a e^{-st} 0 \, dt + \int_a^\infty e^{-st} F(t-a) \, dt \tag{57-5}$$

$$= \int_a^\infty e^{-st} U(t-a) F(t-a) \, dt$$

$$= L \{ U(t-a) F(t-a) \} \tag{57-6}$$

Hence, if $L^{-1} \{ f(s) \} = F(t)$, then

$$L^{-1} \{ e^{-as} f(s) \} = U(t-a) F(t-a)$$

i.e.,

$$L^{-1} \{ e^{-as} f(s) \} = 0, \qquad 0 < t < a$$
$$= F(t-a), \qquad t > a$$

We can now give a simple geometrical interpretation to the function $U(t - \mathbf{a}) F(t - \mathbf{a})$. Let the definition of F(t) be extended such that F(t) is equal to zero for negative values of t. Then the graph of U(t - **a**) F(t - **a**) is actually the graph obtained by shifting the graph of F(t) to the right a distance **a**.

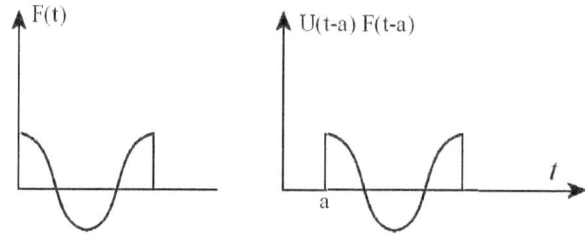

This can also be stated as follows:

If the definition of F(t) be extended such that F(t) is equal to zero for negative values of t, then the effect of shifting the graph of F(t) through a distance **a** to the right is to multiply the transform f(s) by e^{-as}.

Example 35

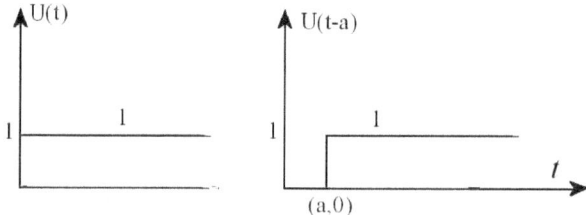

Since the unit step function U(t—a) is actually a shift of U(t) a distance **a** to the right and since

L { U(t) } = 1 / s

Then,

$$L\{U(t-a)\} = \frac{e^{-as}}{s}$$

(58)

Example 36: Inverse Laplace of Power Functions

Since,

$$L^{-1}\left\{\frac{1}{s^4}\right\} = \frac{t^3}{3!}$$

Then,

$$L^{-1}\left\{\frac{e^{-as}}{s^4}\right\} = U(t-a)\frac{(t-a)^3}{3!} \qquad (59)$$

i.e.,

$$L^{-1}\left\{\frac{e^{-as}}{s^4}\right\} = 0, \qquad 0 < t < a$$

$$L^{-1}\left\{\frac{e^{-as}}{s^4}\right\} = \frac{(t-a)^3}{3!}, \qquad t > a$$

Example 37

Since,

$$L^{-1}\left\{\frac{s}{s^2+k^2}\right\} = \cos kt$$

Then,

$$L^{-1}\left\{\frac{s\,e^{-as}}{s^2+k^2}\right\} = U(t-a)\cos k(t-a) \qquad (60)$$

i.e.,

$$L^{-1}\left\{\frac{s\,e^{-as}}{s^2+k^2}\right\} = 0, \quad 0 < t < a$$

$$L^{-1}\left\{\frac{s\,e^{-as}}{s^2+k^2}\right\} = \cos k(t-a), \quad t > a$$

Example 38

Find

$$L^{-1}\left\{\frac{(s+2)\,e^{-\pi s}}{s^2+s+1}\right\} \qquad (61)$$

Solution

First, we will obtain the inverse Laplace transform of the proper fraction as follows

$$L^{-1}\left\{\frac{(s+2)}{s^2+s+1}\right\} = L^{-1}\left\{\frac{s+\dfrac{1}{2}+\dfrac{3}{2}}{\left(s+\dfrac{1}{2}\right)^2+\dfrac{3}{4}}\right\}$$

$$= L^{-1}\left\{\frac{s+\dfrac{1}{2}}{\left(s+\dfrac{1}{2}\right)^2+\dfrac{3}{4}}\right\} + L^{-1}\left\{\frac{\dfrac{3}{2}}{\left(s+\dfrac{1}{2}\right)^2+\dfrac{3}{4}}\right\}$$

$$= e^{-\frac{t}{2}}\cos\left(\frac{\sqrt{3}}{2}t\right) + \sqrt{3}\, e^{-\frac{t}{2}}\sin\left(\frac{\sqrt{3}}{2}t\right)$$

$$= e^{-\frac{t}{2}}\left[\cos\left(\frac{\sqrt{3}}{2}t\right) + \sqrt{3}\sin\left(\frac{\sqrt{3}}{2}t\right)\right] \tag{61-1}$$

Second, we use the second translation property to shift the expression in equation (61-1) by $(-\pi)$.

Therefore,

$$L^{-1}\left\{\frac{(s+2)\, e^{-\pi s}}{s^2+s+1}\right\} = e^{-\frac{t-\pi}{2}}\left[\cos\left(\frac{\sqrt{3}}{2}(t-\pi)\right) + \sqrt{3}\sin\left(\frac{\sqrt{3}}{2}(t-\pi)\right)\right] U(t-\pi) \tag{61-2}$$

i.e.,

$$L^{-1}\left\{\frac{(s+2)\, e^{-\pi s}}{s^2+s+1}\right\} = 0, \qquad 0 < t < \pi$$

$$= e^{-\frac{t-\pi}{2}}\left[\cos\left(\frac{\sqrt{3}}{2}(t-\pi)\right) + \sqrt{3}\sin\left(\frac{\sqrt{3}}{2}(t-\pi)\right)\right], \quad t > \pi$$

Example 39

If

$$F(t) = L^{-1}\left\{\frac{e^{-3s}}{(s+1)^3}\right\} \tag{62}$$

Calculate F(3/2) , F(4)

Solution

First the inverse Laplace transform of the proper fraction is

$$L^{-1}\left\{\frac{1}{(s+1)^3}\right\} = \frac{1}{2}t^2 e^{-t} \tag{62-1}$$

Second, translating the above function by the shift property by (-3) we get

$$F(t) = L^{-1}\left\{\frac{e^{-3s}}{(s+1)^3}\right\} = \frac{1}{2}(t-3)^2\, e^{-(t-3)}\ U(t-3) \tag{62-2}$$

i.e.,

$$F(t) = L^{-1}\left\{\frac{e^{-3s}}{(s+1)^3}\right\} = 0 \qquad 0 < t < 3$$

$$= \frac{1}{2}(t-3)^2\, e^{-(t-3)} \qquad t > 3$$

Therefore,

$$
\begin{array}{ll}
F(3/2) & = 0 \\
F(4) & = (1/2)\,(4\text{-}3)^2\, e^{-(4\text{-}3)} = 1/\,2e
\end{array}
$$

Example 40

Find and sketch the function F(t) for which,

$$F(t) = L^{-1}\left\{\frac{3}{s} - \frac{4e^{-s}}{s^2} + \frac{4e^{-3s}}{s^2}\right\} \tag{63}$$

Solution

Since,

$$L^{-1}\left\{\frac{1}{s^2}\right\} = t \tag{63-1}$$

Then

$$L^{-1}\left\{\frac{e^{-s}}{s^2}\right\} = (t-1)U(t-1) \tag{63-2}$$

And

$$L^{-1}\left\{\frac{e^{-3s}}{s^2}\right\} = (t-3)U(t-3) \tag{63-3}$$

Therefore, substituting from (63-1), (63-2), and (63-3) in (63), we get

$$F(t) = 3 - 4\,(t\text{-}1)\ U(t\text{-}1) + 4\,(t\text{-}3)\ U(t\text{-}3) \tag{63-4}$$

Hence, equation (63 4) implies the following

For $0 < t < 1$, $F(t) = 3$
For $1 < t < 3$, $F(t) = 3\text{-}4(t\text{-}1) = 7\text{ -}4t$
For $t > 3$, $F(t) = 3\text{-}4(t\text{-}1)+4(t\text{-}3) = \text{-}5$

Hence, the graph of F(t) that represent the above three regions is

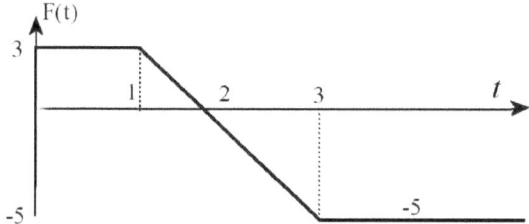

2.3. Application of the shift theorem to the solution of difference and differential equations

Example 41

Solve the first order difference equation

$$Y(t) = Y(t-h) + 1 \quad \text{at} \quad t \geq 0 \tag{64}$$

Given that

$$Y(t) = 0 \quad \text{at.} \quad t < 0$$

Where h being a positive constant.

Solution

The function $Y(t-h)$ is the same as the translated function $U(t-h)\, Y(t-h)$.
Hence transforming the difference equation using the <u>shifting property,</u> we get

$$y(s) = e^{-hs}\, y(s) + \frac{1}{s} \tag{64-1}$$

i.e., arranging we get

$$y(s) = \frac{1}{s(1 - e^{-hs})} \tag{64-2}$$

Expanding the denominator of (64-2) using the Binomial theorem we get

$$y(s) = \frac{1}{s}(1 - e^{-hs})^{-1}$$

$$= \frac{1}{s}(1 - e^{-hs})^{-1} = \frac{1}{s} + \frac{1}{s}e^{-hs} + \frac{1}{s}e^{-2hs} + ... \tag{64-3}$$

Therefore,

$$Y(t) = U(t) + U(t - h) + U(t - 2h) + ... \tag{64-5}$$

i.e.,

$$Y(t) \quad = 1 \qquad 0 < t < h$$

$$= 1+1 = 2 \qquad h < t < 2h$$
$$= 1+1+1 = 3 \qquad 2h < t < 3h \tag{64-6}$$

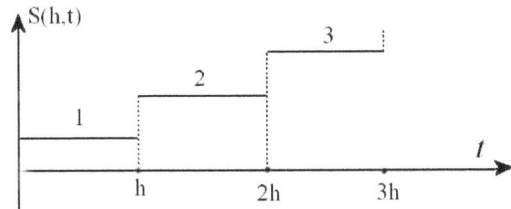

The function Y(t), equation (64-6), is known as a **staircase function,** S(h,t) with a positive run **h** and with a unit rise.

Its Laplace transform in equation (64-3) can be written in alternative form as follows

$$y(s) = \frac{1}{s}(1 - e^{-hs})^{-1} = \frac{e^{\frac{hs}{2}}}{s\left(e^{\frac{hs}{2}} - e^{-\frac{hs}{2}}\right)}$$

$$= \frac{\cosh\left(\frac{hs}{2}\right) + \sin\left(\frac{hs}{2}\right)}{2s \, \sinh\left(\frac{hs}{2}\right)}$$

$$= \frac{1}{2s}\left[1 + \cosh\left(\frac{hs}{2}\right)\right] \tag{64-7}$$

Example 42

Find the function Y(t) which satisfies the second-order difference equation

$$Y(t) - 4Y(t - h) + 4Y(t - 2h) = 1 \tag{65}$$

Given the condition Y(t) = 0, when t < 0 and the right hand side has to be replaced by zero when t < o.

Where, h is a positive constant.

Solution

81

The Laplace transform of equation (65) is

$$y(s) - 4 e^{-hs} y(s) + 4 e^{-2hs} y(s) = \frac{1}{s}$$

(65-1)

Equation (65-1) can be written in the quadratic form as follows

$$y(s)\left(1 - 2 e^{-hs}\right)^2 = \frac{1}{s}$$

Therefore,

$$y(s) = \frac{1}{s\left(1 - 2 e^{-hs}\right)^2} = \frac{1}{s}\left(1 - 2 e^{-hs}\right)^{-2}$$

$$= \frac{1}{s} + \frac{1}{s}(4)e^{-hs} + \frac{(-2)(-3)}{2!s}(4) e^{-2hs} + \ldots$$

$$= \frac{1}{s} + \frac{4}{s}e^{-hs} + \frac{12}{s}e^{-2hs} + \ldots$$

(65-2)

The inverse Laplace transform of (65-2) is

$$Y(t) = U(t) + 4 U(t - h) + 12 U(t - 2h) + \ldots$$

(65-3)

i.e.,

$$
\begin{aligned}
Y(t) \quad &= 1 & 0 < t < h \\
&= 1 + 4 = 5 & h < t < 2h \\
&= 1 + 4 + 12 = 17 & 2h < t < 3h
\end{aligned}
$$

(64-6)

Example 43

Solve the difference - differential equation

$$Y'(t) - 2Y(t-1) = 3$$

(66)

Given that $Y(t) = 0$ when $t \leq 0$ and the constant in the right hand side is to be replaced by zero when $t < o$.

Solution

The Laplace transform of equation (66) is obtained by the use of: (1) transform of derivative; $Y'(t)$, (2) transform of shifted function; $Y(t-1)$, and (3) transform of constant; 3.

Thus,

$$s\,y(s) - 2 e^{-s} y(s) = \frac{3}{s}$$

(66-1)

$$\left(s - 2 e^{-s}\right)y(s) = \frac{3}{s}$$

Therefore,

$$y(s) = \frac{3}{s\left(s - 2 e^{-s}\right)}$$

(66-2)

The denominator of (66-2) can be expanded by the use of the Binomial theorem as follows

$$y(s) = \frac{3}{s\left(s - 2\,e^{-s}\right)} = \frac{3}{s^2\left(1 - \dfrac{2\,e^{-s}}{2}\right)} = \frac{3}{s^2}\left(1 - \frac{2\,e^{-s}}{2}\right)^{-1}$$

$$= \frac{3}{s^2}\left(1 + \frac{2\,e^{-s}}{s} + \frac{4\,e^{-2s}}{s^2} + \dots\right)$$

$$= \frac{3}{s^2} + \frac{6\,e^{-s}}{s^3} + \frac{12\,e^{-2s}}{s^4} + \dots$$

Therefore,

$$Y(t) = 3t + \frac{6}{2}(t-1)^2\,U(t-1) + \frac{12}{3!}(t-2)^3\,U(t-2) + \dots \qquad (66\text{-}3)$$

i.e.,

$$
\begin{aligned}
Y(t) \quad &= 3t && 0 < t < 1 \\
&= 3t + 3(t\text{-}1)^2 && 1 < t < 2 \\
&= 3t + 3(t\text{-}1)^2 + 2(t\text{-}2)^3 && 2 < t < 3
\end{aligned}
\qquad (66\text{-}4)
$$

Example 44

Compute $y(\pi/2)$ and $y(2 + \pi/2)$ for the function $y(x)$ which satisfies the boundary-value problem

$$y''(x) + y(x) = (x\text{-}2)\,U(x\text{-}2) \qquad (67)$$

Given that

$$y(0) = y'(0) = 0$$

Solution

Equation (67) implies the following

$$
\begin{aligned}
(x\text{-}2)\,U(x\text{-}2) \quad &= 0 && 0 < x < 2 \\
&= x\text{-}2 && x > 2
\end{aligned}
\qquad (67\text{-}1)
$$

The Laplace transforming of the differential equation (67) gives

$$\left(s^2 + 1\right)y(s) = \frac{e^{-2s}}{s^2} \qquad (67\text{-}2)$$

Therefore,

$$y(s) = \frac{e^{-2s}}{s^2\left(s^2 + 1\right)} \qquad (67\text{-}3)$$

Writing equation (67-2) in its partial fractions gives

$$y(s) = e^{-2s}\left(\frac{A}{s^2} + \frac{B}{s^2 + 1}\right) \qquad (67\text{-}4)$$

Where,

$$A(s^2 + 1) + B\,s^2 = 1 \qquad (67\text{-}5)$$

Therefore,

A + B = 0
A = 1
Thus
B = -1

Therefore, equation (67-4) becomes

$$y(s) = e^{-2s}\left(\frac{1}{s^2} - \frac{1}{s^2 + 1}\right)$$

(67-6)

The inverse Laplace transform of (67-6) is

$$Y(t) = [(x - 2) - \sin(x - 2)]\, U(t - 2)$$

(67-7)

Therefore,

$$
\begin{aligned}
y(x) \quad &= 0 && 0 < x < 2 \\
&= (x-2) - \sin(x-2) && x > 2
\end{aligned}
$$

(67-8)

Thus, for y(π /2) and y(2 + π /2) we get

$$
\begin{aligned}
y(\pi /2) \quad &= 0 && 0 < x < 2 \\
y(2 + \pi /2) &= (2 + \pi /2\text{-}2) - \sin(2 + \pi /2\text{-}2) \\
&= (\pi /2) - \sin(\pi /2) \\
&= (\pi /2) - 1
\end{aligned}
$$

Example 45

Solve the equation

$$x'' (t) + 4 x(t) - \psi (t)$$

(68)

Given that

$$x(0) = 1$$
$$x' (0) = 0$$

Where ψ (t) is defined as

$$
\begin{aligned}
\psi (t) &= 4t && 0 \le t \le 1 \\
&= 4 && t > 1
\end{aligned}
$$

Solution

We write

$$\psi (t) - 4\,t - 4(t\text{-}1)\, U\,(t\text{-}1)$$

(68-1)

The Laplace transform of (68-1) is

$$L\{\psi(t)\} = \frac{4}{s^2} - \frac{4}{s^2}e^{-s}$$

(68-2)

84

The Laplace transform of (68) is

$$L\{x''(t)+4x(t)\}=s^2 x(s)-s+4 x(s) \tag{68-3}$$

Therefore, from (68-2), (68-3), and (68), we get

$$\frac{4}{s^2}-\frac{4}{s^2}e^{-s}=s^2 x(s)-s+4 x(s) \tag{68-4}$$

Thus,

$$x(s)=\frac{s}{s^2+4}+\frac{4}{s^2(s^2+4)}-\frac{4}{s^2(s^2+4)}e^{-s} \tag{68-5}$$

$$=\frac{s}{s^2+4}+\frac{4}{s^2(s^2+4)}(1-e^{-s})$$

$$=\frac{s}{s^2+4}+\left[\frac{1}{s^2}-\frac{1}{s^2+4}\right](1-e^{-s}) \tag{68-6}$$

The inverse Laplace transform of (68-6) is

$$x(t)=\cos 2t + t - (1/2)\sin 2t - [(t-1)-(1/2)\sin 2(t-1)]\, U(t-1) \tag{68-7}$$

2.4. The unit impulse function

Consider the function F(t) defined by

$$
\begin{array}{lll}
F(t) & =1/\varepsilon & 0<t<\varepsilon \\
& =0 & t>\varepsilon
\end{array} \tag{69}
$$

Thus, the area under the function is: $(1/\varepsilon)\ \varepsilon=1$

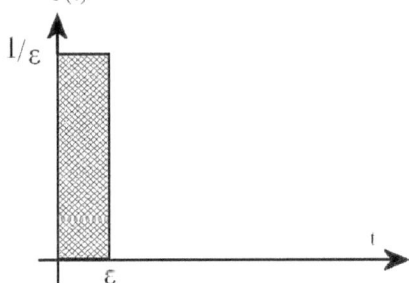

Its Laplace transform of the unit function, equation (69), is

$$\int_0^\varepsilon e^{-st}\frac{1}{\varepsilon}dt=-\frac{1}{\varepsilon}\left[\frac{e^{-st}}{-s}\right]_0^\varepsilon=\frac{1}{\varepsilon s}(1-e^{-\varepsilon s}) \tag{69-1}$$

85

Now, let us take the limiting value of the transform when $\varepsilon \to 0$

$$\lim_{\varepsilon \to 0} \frac{1-e^{-\varepsilon s}}{\varepsilon s} = \lim_{\varepsilon \to 0} \frac{s\, e^{-\varepsilon s}}{s} = 1 \qquad (69\text{-}2)$$

The limiting function of F(t) when $\varepsilon \to 0$ is known as the **unit-impulse function** or the **Dirac delta** function $\delta(t)$.

Thus ,

$$L\{\,\delta(t)\,\} = 1 \qquad (70)$$

If the function be shifted a distance T along the positive t-axis, then according to the second shift theorem its transform is

$$L\{\delta(t-T)\} = e^{-Ts} \qquad (70\text{-}1)$$

We shall now show that the derivative of the **unit-step function** is the **impulse function**. For, consider the function

$$\frac{1}{\varepsilon}U(t) - \frac{1}{\varepsilon}U(t-\varepsilon) \qquad (70\text{-}2)$$

Its transform is

$$L\left\{\frac{1}{\varepsilon}U(t) - \frac{1}{\varepsilon}U(t-\varepsilon)\right\} = \frac{1}{\varepsilon}\left(\frac{1}{s} - \frac{1}{s}e^{-\varepsilon s}\right) = \frac{1}{\varepsilon s}\left(1 - e^{-\varepsilon s}\right) \qquad (70\text{-}3)$$

And, as $\varepsilon \to 0$, the expression (69-2) tends formally to U'(t) and its transform (70-1) tends to 1.

Thus,

$$L\{U'(t)\} = L\{\delta(t)\} = 1 \qquad (70\text{-}4)$$

And

$$L\{\,U'(t-T)\,\} = L\{\delta(t-T)\} = e^{-Ts} \qquad (70\text{-}5)$$

Example 46

A series circuit of resistance, R, and inductance, L, is connected to a generator, which delivers an impulse voltage $V_o\,\delta(t)$, which is impressed upon the circuit at zero time. Find the current in the circuit.

Solution

Ohm's law for combined resistance drain (RI) and inductance impulse (L dI/dt) resulting from voltage impulse $V_o\,\delta(t)$, is written as follows.

$$L\frac{dI}{dt} + RI = V_o\,\delta(t) \qquad (71)$$

86

The Laplace transform of (71) is obtained by: (1) using the property of Laplace transform that implies the multiplication of a function by s is equal to the derivative of the function, (2) Laplace transform of $\delta(t)$ is equal to unity. We get

$$L\,s\,i(s) + R\,i(s) = V_o \qquad\qquad (71\text{-}1)$$

Initially, $I(0) = 0$, because the voltage impulse was applied on an inactive circuit.

Thus,

$$(Ls + R)i(s) = V_o$$

Therefore,

$$i(s) = \frac{V_o}{Ls + R} = \frac{V_o}{L} \cdot \frac{1}{s + \dfrac{R}{L}} \qquad\qquad (71\text{-}2)$$

The inverse Laplace transform of (71-2) gives the temporal function of the electric current $I(t)$ as

$$I(t) = \frac{V_o}{L} \cdot e^{-\frac{R}{L}t} \qquad\qquad (71\text{-}3)$$

Note:
An impulsive voltage means a large voltage acting for a very short interval of time but the time integral is finite.

2.5. The unit doublet

Consider the function $F(t)$ defined by

$$F(t) = \frac{1}{\varepsilon^2} \qquad\qquad 0 < t < \varepsilon$$

$$= -\frac{1}{\varepsilon^2} \qquad\qquad \varepsilon < t < 2\varepsilon$$

$$= 0 \qquad\qquad t > 2\varepsilon \qquad\qquad (72)$$

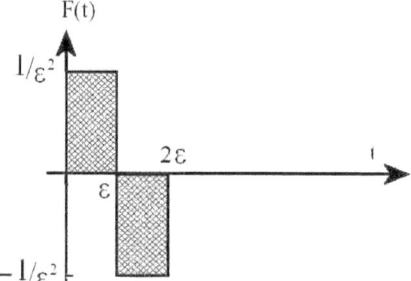

The Laplace transform of (72) is

$$L\{F(t)\} = \int_0^\varepsilon \frac{1}{\varepsilon^2} e^{-st} dt - \int_\varepsilon^{2\varepsilon} \frac{1}{\varepsilon^2} e^{-st} dt \qquad (72\text{-}1)$$

$$= \frac{1}{\varepsilon^2 s}\left[1 - 2e^{-s\varepsilon} + e^{-2s\varepsilon}\right]$$

$$= \frac{1}{\varepsilon^2 s}\left(1 - e^{-s\varepsilon}\right)$$

Let us find the limiting value of this transform when $\varepsilon \to \infty$

$$\lim_{\varepsilon \to 0} \frac{1 - 2e^{-s\varepsilon} + e^{-2s\varepsilon}}{\varepsilon^2 s} = \lim_{\varepsilon \to 0} \frac{\dfrac{d^2}{d\varepsilon^2}\left(1 - 2e^{-s\varepsilon} + e^{-2s\varepsilon}\right)}{\dfrac{d^2}{d\varepsilon^2}\left(\varepsilon^2 s\right)}$$

$$= \lim_{\varepsilon \to 0} \frac{-2s^2 e^{-s\varepsilon} + 4s^2 e^{-2s\varepsilon}}{2s}$$

$$= \frac{2s^2}{2s} = s \qquad (72\text{-}2)$$

The limiting function of F(t) when $\varepsilon \to \infty$ is known as the **unit doublet** and its transform is s.

If the **unit doublet** is shifted along the positive t-axis a distance T, then its transform is

$s\, e^{-Ts}$.

We shall now show that the **derivative of the unit impulse function** is the unit doublet.

For, consider the function

88

$$\frac{1}{\varepsilon}\delta(t) - \frac{1}{\varepsilon}\delta(t - \varepsilon) \qquad (72\text{-}3)$$

Where, ε may he made as small as we please.

The Laplace transform of (72-3) is obtained by: (1) The unity property of the $\delta(t)$ and (2) shift property of $\delta(t - \varepsilon)$; s $e^{-\varepsilon\, s}$.

Therefore,

$$L\left\{\frac{1}{\varepsilon}\delta(t) - \frac{1}{\varepsilon}\delta(t - \varepsilon)\right\} = \frac{1}{\varepsilon} - \frac{1}{\varepsilon}e^{-\varepsilon s} = \frac{1}{\varepsilon}\left(1 - e^{-\varepsilon s}\right) \qquad (72\text{-}4)$$

And, as $\varepsilon \to \infty$ the expression (72-3) tends formally to $\delta'(t)$ and its transform (72-4) tends to s .

Then

$$L\{\delta'(t)\} = L\left\{\frac{1}{\varepsilon}\delta(t) - \frac{1}{\varepsilon}\delta(t - \varepsilon)\right\} = s \qquad (72\text{-}5)$$

And $\qquad L\{\delta'(t - T)\} = s\, e^{-Ts} \qquad (72\text{-}5)$

From (70-4),
$$L\{U'(t)\} = L\{\delta(t)\} = 1$$

Thus,
$$L\{U''(t)\} = s \qquad (72\text{-}6)$$

Also, from (70-5),

$$L\{U'(t - T)\} = L\{\delta(t - T)\} = e^{-Ts}$$

Thus,

$$L\{U''(t - T)\} = s\, e^{-Ts} \qquad (72\text{-}7)$$

.
2.6. The behavior of f(s) as $s \to \infty$

Let $F(t)$ be a function which is sectionally continuous in every finite interval $0 \le t \le T$. And, let it be of the order of $e^{\alpha_0 t}$ as $t \to \infty$.

i.e.,
$$|F(t)| < M e^{\alpha_0 t}$$

Consider $s \ge \alpha$, where $\alpha \ge \alpha_0$

Then,
$$|e^{-st} F(t)| < e^{-\alpha t} M e^{\alpha_0 t} = M e^{-(\alpha - \alpha_0)t}$$

89

Therefore,

$$| f(s) | = | L \{F(t)\} | = \left| \int_0^\infty e^{-st}F(t)dt \right|$$

$$\leq \int_0^\infty \left| e^{-st}F(t) \right| dt$$

$$\leq \int_0^\infty M e^{-(\alpha-\alpha_o)t} \, dt$$

$$= M \left[\frac{e^{-(\alpha-\alpha_o)t}}{\alpha_o - \alpha} \right]_0^\infty = = \frac{M}{\alpha_o - \alpha}, \quad s \geq \alpha$$

Put $\alpha = s$,

Therefore,

$$| f(s) | < \frac{M}{\alpha_o - \alpha} \rightarrow 0 \text{ as } s \rightarrow \infty$$

Therefore,

$$f(s) \rightarrow 0 \text{ as } s \rightarrow \infty \tag{73}$$

2.7. Initial-value theorem

To prove that

$$\lim_{t \to 0} F(t) = \lim_{s \to \infty} s\, f(s)$$

$$L\{F'(t)\} = \int_0^\infty e^{-st}F'(t)dt = s\, f(s) - F(0)$$

Now, if F'(t) is sectionally continuous and of exponential order then according to the theorem in 2.6, we get

$$\lim_{s \to \infty} \int_0^\infty e^{-st}F'(t)dt = 0$$

Taking the limit in (73), assuming F(t) to be continuous at t = 0, we get

$$0 = \lim_{s \to \infty} s\, f(s) - F(0)$$

i.e.,

$$\lim_{s \to \infty} s\, f(s) - F(0) = \lim_{t \to 0} F(t) \tag{74}$$

2.8. Final-value theorem

To prove that

$$\lim_{t \to \infty} F(t) = \lim_{s \to 0} s\, f(s)$$

$$L\left\{F'(t)\right\} = \int_0^\infty e^{-st}F'(t)dt = s\,f(s) - F(0) \qquad (75)$$

The limit of the left hand side in equation (75) as s $\rightarrow$ 0 is

$$\lim_{s \to 0} \int_0^\infty e^{-st}F'(t)dt = \int_0^\infty F'(t)dt = \left[F(t)\right]_0^\infty$$

$$= \lim_{t \to \infty} F(t) - F(0) \qquad (75\text{-}1)$$

The limit of right hand side of (75) as s $\rightarrow$ 0 is

$$\lim_{s \to 0} s\,f(s) - F(0) \qquad (75\text{-}2)$$

From (75-1) and (75-2), we get

$$\lim_{t \to \infty} F(t) - F(0) = \lim_{s \to 0} s\,f(s) - F(0)$$

i.e.,

$$\lim_{t \to \infty} F(t) = \lim_{s \to 0} s\,f(s) \qquad (75\text{-}3)$$

Example 47

Let $F(t) = 2\,e^{-3t}$, then

$$f(s) = \frac{2}{s+3}$$

By the initial-value theorem, we have

$$\lim_{t \to 0} 2\,e^{-3t} = \lim_{s \to \infty} \frac{2s}{s+3} = 2$$

And, by the final-value theorem, we have

$$\lim_{t \to \infty} 2\,e^{-3t} = \lim_{s \to 0} \frac{2s}{s+3} = 0$$

2.9. Differentiation of transform

Let F(t) be a sectionally continuous function in every finite interval $0 \le t \le T$ and let it be of the order of $e^{\alpha t}$ as $t \rightarrow \infty$.

Then if $s > 0$, we have

$$L\left\{F(t)\right\} = \int_0^\infty e^{-st}F(t)\,dt = f(s) \qquad (76)$$

Differentiate with respect to s

$$\frac{\partial}{\partial s} L\{F(t)\} = \int_0^\infty \frac{\partial}{\partial s}\{e^{-st}F(t)\}dt = \frac{\partial}{\partial s}f(s)$$

$$= \int_0^\infty -t\,e^{-st}\,F(t)\,dt$$

$$= L\{-t\,F(t)\} = f'(s) \tag{76-1}$$

Similarly,

$$L\{(-t)^2\,F(t)\} = f''(s) \tag{76-2}$$

Therefore, in general, we get

$$L\{(-t)^n\,F(t)\} = f^{(n)}(s) \tag{76-3}$$

Hence, the differentiation of the transform of a function *with respect to* "s" corresponds to the multiplication of the function by $(-t)$.

e.g.,

$$L\{\sin kt\} = \frac{k}{s^2+k^2}, \qquad s > 0 \tag{77}$$

Therefore,

$$L\{-t\,\sin kt\} = \frac{-2sk}{\left(s^2+k^2\right)^2} \tag{77-1}$$

Then

$$L^{-1}\left\{\frac{s}{\left(s^2+k^2\right)^2}\right\} = \frac{1}{2k}t\,\sin kt \tag{77-2}$$

Similarly,

$$L\{\cos kt\} = \frac{s}{s^2+k^2}, \tag{78}$$

Therefore,

$$L\{-t\,\cos kt\} = \frac{s^2+k^2-2s^2}{\left(s^2+k^2\right)^2} = \frac{k^2-s^2}{\left(s^2+k^2\right)^2}$$

$$= \frac{A}{s^2+k^2} + \frac{B}{\left(s^2+k^2\right)^2} \tag{78-1}$$

Where, $A(s^2 + k^2) + B = k^2 - s^2$
Thus,
$A = -1$

$A k^2 + B = k^2$

$B = 2 k^2$

Thus, equation (78-1) becomes

$$L\left\{ -t \cos kt \right\} = -\frac{1}{s^2 + k^2} + \frac{2 k^2}{\left(s^2 + k^2\right)^2}$$ (78-2)

Replace $\dfrac{1}{s^2 + k^2} = L\left\{ \dfrac{\sin kt}{k} \right\}$.

Equation (78-2) becomes

$$L\left\{ t \cos kt - \frac{1}{k}\sin kt \right\} = -\frac{2 k^2}{\left(s^2 + k^2\right)^2}$$

Therefore,

$$L^{-1}\left\{ \frac{1}{\left(s^2 + k^2\right)^2} \right\} = \frac{1}{2k^2}\left(\sin kt - kt \cos kt \right)$$

2.9. Differentiation of transform

Let F(t) be a sectionally continuous function in every finite interval $0 \leq t \leq T$ and let it be of the order of $e^{\alpha t}$ as $t \to \infty$.

Then if s > 0, we have

$$L\left\{ F(t) \right\} = \int_0^\infty e^{-st} F(t)\, dt = f(s)$$ (79)

Differentiate with respect to s

$$\frac{\partial}{\partial s} L\left\{ F(t) \right\} = \int_0^\infty \frac{\partial}{\partial s}\left\{ e^{-st} F(t) \right\} dt = \frac{\partial}{\partial s} f(s)$$

$$= \int_0^\infty -t e^{-st} F(t)\, dt$$

$$= L\left\{ -t F(t) \right\} = f'(s)$$ (79-1)

Similarly,

$$L\left\{ (-t)^2 F(t) \right\} = f''(s)$$ (79-2)

Therefore, in general, we get

$$L\left\{ (-t)^n F(t) \right\} = f^{(n)}(s)$$ (79-3)

Hence, the differentiation of the transform of a function *with respect to* "s" corresponds to the multiplication of the function by $(-t)$.

93

e.g.,

$$L\{\sin kt\} = \frac{k}{s^2 + k^2}, \qquad\qquad s > 0 \qquad\qquad (80)$$

Therefore,

$$L\{-t \sin kt\} = \frac{-2sk}{(s^2 + k^2)^2} \qquad\qquad (80\text{-}1)$$

Then

$$L^{-1}\left\{\frac{s}{(s^2 + k^2)^2}\right\} = \frac{1}{2k} t \sin kt \qquad\qquad (80\text{-}2)$$

Similarly,

$$L\{\cos kt\} = \frac{s}{s^2 + k^2}, \qquad\qquad (81)$$

Therefore,

$$L\{-t \cos kt\} = \frac{s^2 + k^2 - 2s^2}{(s^2 + k^2)^2} = \frac{k^2 - s^2}{(s^2 + k^2)^2}$$

$$= \frac{A}{s^2 + k^2} + \frac{B}{(s^2 + k^2)^2} \qquad\qquad (81\text{-}1)$$

Where, $A(s^2 + k^2) + B = k^2 - s^2$
Thus,
$A = -1$

$A k^2 + B = k^2$
$B = 2 k^2$

Thus, equation (81-1) becomes

$$L\{-t \cos kt\} = -\frac{1}{s^2 + k^2} + \frac{2 k^2}{(s^2 + k^2)^2} \qquad\qquad (81\text{-}2)$$

Replace $\dfrac{1}{s^2 + k^2} = L\left\{\dfrac{\sin kt}{k}\right\}$.

Equation (81-2) becomes

$$L\left\{t \cos kt - \frac{1}{k}\sin kt\right\} = -\frac{2 k^2}{(s^2 + k^2)^2}$$

Therefore,

$$L^{-1}\left\{\frac{1}{(s^2 + k^2)^2}\right\} = \frac{1}{2k^2}(\sin kt - kt \cos kt)$$

2.10. Application of the differentiation of Laplace transform to the solution of linear differential equations with coefficients as polynomials in t

Example 48

Find a solution of the differential equation

$$t\frac{d^2x}{dt^2} + t\frac{dx}{dt} + x = 0 \tag{83}$$

Given that which satisfies $x(0) = 0$ and $dx/dt = 1$ when $t = 0$.

Solution

The Laplace transformation of (83) is obtained in the usual manner using:

(1) The property that differential derivative of the function corresponds to multiplication by s on the Laplace transform. [Equations (24), (25), and (26)]

$$L\{F''(t)\} = s^2 \ f(s) - s F(0) - F'(0) \tag{83-1}$$

(2) The property of multiplication of function by t corresponds to differentiation of the Laplace transform with respect to s. [Equation (27-5)]

$$L\left\{ \int_0^t F(\tau)\, d\tau \right\} = \frac{1}{s}\left(f(s) + G(0) \right) \tag{83-2}$$

From (83-1) and (83-2), the Laplace transformation of (83) is

$$-\frac{d}{ds}\left(s^2 \ x(s) - s \ x(0) - x'(0) \right) - \frac{d}{ds}\left(s \ x(s) - x(0) \right) + x(s) = 0 \tag{83-3}$$

In (83-3), the derivative (d/ds) corresponds to the integration of the function or the term (t) in (83). The multiplication by (s) corresponds to the differentiation of the function.

Substitute in (83) by $x(0) = 0$ and $dx/dt = 1$ when $t = 0$, therefore,

$$\frac{d}{ds}\left(s^2 \ x(s) - 0 - 1 \right) - \frac{d}{ds}\left(s \ x(s) - 0 \right) + x(s) - 0$$

$$-s^2 \frac{dx(s)}{ds} - 2s \ x(s) - s\frac{dx(s)}{ds} - x(s) + x(s) = 0$$

$$s\left[(s+1)\frac{dx(s)}{ds} + 2 \ x(s) \right] = 0 \tag{83-4}$$

Diving (83-4) by $s(1 + s) x(s)$ and integrating we get

95

$$\int \frac{dx(s)}{x(s)} + 2 \int \frac{ds}{1+s} = 0 \tag{83-5}$$

Thus,

$$\log x(s) + 2 \log (1+s) = \log C$$

Therefore,

$$x(s) = \frac{C}{(1+s)^2} \tag{83-6}$$

The inverse Laplace transform of (83-6) is

$$x(t) = L^{-1}\{x(s)\} = L^{-1}\left\{\frac{C}{(1+s)^2}\right\}$$

$$= C \ t \ e^{-t} \tag{83-7}$$

The constant C is determined by putting $dx/dt = 1$ when $t = 0$, which gives $C = 1$
Hence,

$$x(t) = t \ e^{-t}$$

Example 49

Consider the Bessel differential equation of order α as follows

$$x^2 \frac{d^2 y}{dx^2} + x \frac{dy}{dx} + \left(x^2 - \alpha^2\right) y = 0 \tag{84}$$

Find the solution of Bessel's differential equation of zero order ($\alpha = 0$)

$$t \ Y''(t) + Y'(t) + t \ Y(t) = 0 \tag{84-1}$$

Which has a Laplace transform and which satisfies $Y(0) = 1$

Solution

The Laplace transform of equation (84-1) gives

$$-\frac{d}{ds}\left(s^2 \ y(s) - s - Y'(0)\right) + \left(s \ y(s) - Y(0)\right) - \frac{d}{ds}y(s) = 0$$

Therefore,

$$-2 s \ y - s^2 \frac{dy(s)}{ds} + 1 + s \ y(s) - 0 - \frac{dy(s)}{ds} = 0 \tag{84-2}$$

Which gives

$$\left(s^2 + 1\right) \frac{dy(s)}{ds} + s \ y(s) = 0$$

96

Therefore,

$$\int \frac{dy(s)}{y(s)} = -\int \frac{s\ ds}{s^2 + 1}$$

$$\log\ y = -\frac{1}{2}\ \log\ (s^2 + 1) + \log\ C$$

$$\log\ y = -\ \log\ \sqrt{(s^2 + 1)} + \log\ C = \log\frac{C}{\sqrt{(s^2 + 1)}}$$

$$y(s) = \frac{C}{\sqrt{(s^2 + 1)}} \tag{84-3}$$

Now, by the initial-value theorem gives

$$\lim_{t \to 0}\ Y(t) = \lim_{s \to \infty}\ s\ y(s) \tag{84-4}$$

From the given condition $Y(0) = 1$, therefore, (84-4) becomes

$$Y(0) = 1 = \lim_{t \to 0}\ Y(t) = \lim_{s \to \infty}\ s\ y(s) \tag{84-4a}$$

From (84-3) and (84-4a), we get

$$\lim_{s \to \infty}\ s\ y(s) = \lim_{s \to \infty}\ \frac{s\ C}{\sqrt{(s^2 + 1)}} = 1 \tag{84-4b}$$

$$\lim_{s \to \infty}\ \frac{C}{\sqrt{\left(1 + \frac{1}{s^2}\right)}} = 1, \text{ which implies } C = 1 \tag{84-5}$$

Thus, C can be determined from equation (84-3) as follows:

$$y(s) = \frac{C}{\sqrt{(s^2 + 1)}} = \frac{C}{s} \cdot \frac{1}{\sqrt{1 + \frac{1}{s^2}}}$$

$$= \frac{C}{s}\left(1 + \frac{1}{s^2}\right)^{-\frac{1}{2}} = \frac{C}{s} \cdot \left[1 - \frac{1}{2} \cdot \frac{1}{s^2} - \left(-\frac{1}{2}\right)\left(\frac{3}{2}\right) \cdot \frac{1}{2!} \cdot \frac{1}{s^4} \cdots \right]$$

$$= \frac{C}{s} \cdot \left[1 - \frac{1}{2} \cdot \frac{1}{s^2} + \frac{1.3}{2^2.2!} \cdot \frac{1}{s^4} - \cdots \right]$$

$$= C \cdot \left[\frac{1}{s} - \frac{1}{2} \cdot \frac{1}{s^3} + \frac{1.3}{2^2.2!} \cdot \frac{1}{s^5} - \cdots \right] \tag{84-6}$$

The inverse Laplace transform of (84-6) is

$$Y(t) = L^{-1}\left\{C.\left[\frac{1}{s} - \frac{1}{2}.\frac{1}{s^3} + \frac{1.3}{2^2.2!}.\frac{1}{s^5} - ..\right]\right\} = C.\left[1 - \frac{t^2}{2^2} + \frac{1.3}{2^2.2!}.\frac{t^4}{4!} - ..\right] \qquad (84\text{-}7)$$

And, since $Y(0) = 1$, therefore $C = 1$ and we get Bessel function of zero order of the first kind, namely,

$$J_0(t) = 1 - \frac{t^2}{2^2} + \frac{t^4}{2^2.4^2.} - \frac{t^6}{2^2.4^2.6^2} + .. \qquad (84\text{-}8)$$

Corollary 1

From (84-3)

$$J_0(t) = L^{-1}\ \{y(s)\} = L^{-1}\left\{\frac{1}{\sqrt{(s^2 + 1)}}\right\}$$

And

$$L\{J_0(t)\} = \frac{1}{\sqrt{(s^2 + 1)}}, \qquad s > 0 \qquad (85)$$

Corollary 2

$$L\{J_0(at)\} = \frac{1}{a}.\frac{1}{\sqrt{\left(\frac{s}{a}\right)^2 + 1}} = \frac{1}{\sqrt{s^2 + a^2}}$$

Corollary 3

Since, $J_0'(t) = -J_1(t)$

Then, $L\ \{J_0'(t)\} = L\ \{-J_1(t)\}$

From the property of Laplace transform of derivatives we get

$$s\ L\ \{J_0(t)\} - J_0(0) = L\ \{-J_1(t)\} \qquad (86)$$

From equation (84-8), $J_0(0) = 1$ and from equation (85), $L\{J_0(t)\} = \dfrac{1}{\sqrt{(s^2 + 1)}}$, equation (86)

becomes

$$\frac{s}{\sqrt{s^2 + 1}} - 1 = -L\ \{J_1(t)\} \qquad (86\text{-}1)$$

Therefore,

$$L\{J_1(t)\} = 1 - \frac{s}{\sqrt{s^2+1}} = \frac{\sqrt{s^2+1} - s}{\sqrt{s^2+1}} \qquad (86\text{-}2)$$

Multiplying the denominator and numerator of (86-2) by $\sqrt{s^2+1} + s$, we get

$$L\{J_1(t)\} = \frac{s^2+1-s^2}{\left[\sqrt{s^2+1} + s\right]\sqrt{s^2+1}} \qquad (86\text{-}3)$$

Example 50

Find the Laplace transform of $\sin\sqrt{t}$

Solution

Put

$$Y(t) = \sin\sqrt{t} \qquad (87)$$

By differentiating twice, we get

$$Y'(t) = \frac{1}{2\sqrt{t}}\cos\sqrt{t} \qquad (87\text{-}1)$$

$$Y''(t) = -\left(\frac{1}{2\sqrt{t}}\right)^2 \sin\sqrt{t} - \frac{1}{4\sqrt{t^3}}\cos\sqrt{t} \qquad (87\text{-}2)$$

Substituting from (87) and (87-1) in (87-2), we get

$$Y''(t) = -\left(\frac{1}{4t}\right)Y(t) - \frac{1}{2t}Y'(t) \qquad (87\text{-}3)$$

Thus,

$$4t\,Y''(t) + 2\,Y'(t) + Y(t) = 0 \qquad (87\text{-}4)$$

The Laplace transform of (87-4) is

$$-4\frac{d}{ds}\left[s^2 y(s) - sY(0) - Y'(0)\right] + 2\left[s^2 y(s) - sY(0)\right] + y(s) = 0 \qquad (87\text{-}5)$$

Performing the differentiation in (87-5) and putting $Y(0) = 0$ gives

$$-4s^2 \frac{dy(s)}{ds} - 8s\ y(s) + 2s\ y(s) + y(s) = 0$$

$$4s^2 \frac{dy(s)}{ds} + (6s - 1)\ y(s) = 0 \tag{87-6}$$

Putting equation (87-6) in integration form follows

$$\int \frac{dy(s)}{y(s)} = -\int \frac{(6s - 1)ds}{4s^2} \tag{87-7}$$

Thus,

$$\log\ y(s) = -\frac{3}{2}\ \log s - \frac{1}{4s} + C$$

$$\log\ y(s) + \log\ s^{\frac{3}{2}} = \log\left(y(s)\ s^{\frac{3}{2}}\right) = -\frac{1}{4s} + C$$

$$y(s)\ = C\ s^{-\frac{3}{2}} e^{-\frac{1}{4s}} \tag{87-8}$$

For **small values** of t, $\sin\sqrt{t} \approx \sqrt{t}$, the Laplace transform of $\sqrt{t}$ has been proved earlier and is given as

$$L\left\{\ \sqrt{t}\ \right\} = \frac{\sqrt{\pi}}{2s^{\frac{3}{2}}} \tag{87-9}$$

For large values of s, equation (87-8) gives

$$y(s)\ = C\ s^{-\frac{3}{2}} e^0 = C\ s^{-\frac{3}{2}} \tag{87-10}$$

Comparing (87-9) and (87-10), we notice that

$$C = \frac{\sqrt{\pi}}{2} \tag{87-11}$$

Alternative method

Expand $\sin\sqrt{t}$ as follows

$$\sin\sqrt{t} = \sqrt{t} - \frac{\left(\sqrt{t}\right)^3}{3!} + \frac{\left(\sqrt{t}\right)^5}{5!} - \dots \tag{88}$$

$$= t^{\frac{1}{2}} - \frac{t^{\frac{3}{2}}}{3!} + \frac{t^{\frac{5}{2}}}{5!} - \dots \tag{88-1}$$

Performing the Laplace transform of (88-1) gives

$$L\left\{\ \sin\sqrt{t}\ \right\} = \frac{\Gamma\left(\frac{3}{2}\right)}{s^{\frac{3}{2}}} - \frac{\Gamma\left(\frac{5}{2}\right)}{3!s^{\frac{5}{2}}} + \frac{\Gamma\left(\frac{7}{2}\right)}{5!s^{\frac{7}{2}}} \dots \tag{88-2}$$

100

$$= \frac{\sqrt{\pi}}{2s^{3/2}}\left[1 - \left(1/2^2 s\right) + \frac{\left(1/2^2 s\right)^2}{2!} - ....\right] \qquad (88\text{-}3)$$

$$= \frac{\sqrt{\pi}}{2s^{3/2}} e^{-\left(1/2^2 s\right)} = \frac{\sqrt{\pi}}{2s^{3/2}} e^{-(1/4s)} \qquad (88\text{-}4)$$

Thus, equation (88-4) is the same as (87-8) with (87-11) being used for C

2.11. Integration of transforms

Let F(t) be sectionally continuous of the order of $e^{\alpha t}$ as t → ∞ and let x > a, then

$$f(x) = \int_0^\infty e^{-xt} F(t) dt \qquad (89)$$

Now, let s > α

And, suppose that $\lim\limits_{t \to +0} \dfrac{F(t)}{t}$ exists.

Then

$$\int_s^\infty f(x) dx = \int_s^\infty \left\{ \int_0^\infty e^{-xt} F(t) dt \right\} dx \qquad (90)$$

Assuming that we can invert the order of integration, which is true in this case, we have

$$\int_s^\infty f(x) dx = \int_0^\infty \left\{ \int_s^\infty e^{-xt} F(t) dx \right\} dt$$

$$= \int_0^\infty F(t) \left[\frac{e^{-xt}}{-t} \right]_s^\infty dt$$

$$= \int_0^\infty \frac{F(t) e^{-xt}}{t} dt$$

$$= \int_0^\infty e^{-xt} \left\{ \frac{F(t)}{t} \right\} dt = L\left\{ \frac{F(t)}{t} \right\} \qquad (90\text{-}1)$$

And, since L {F(t)} = f(s), therefore,

$$L\left\{ \frac{F(t)}{t} \right\} = \int_s^\infty f(x) dx \qquad (91)$$

Hence,

The division of the object function by t corresponds to the integration of the transform of that function from s to ∞.

Example 51

Prove that

$$L\left\{\frac{1-\cos\ kt}{t}\right\} = \frac{1}{2}\log\left(1+\frac{k^2}{s^2}\right), \qquad s > k > 0 \tag{92}$$

Solution

We first find the Laplace transform of the function, 1 −cos kt, as follows

$$L\left\{1-\cos\ kt\right\} = \frac{1}{s} - \frac{s}{s^2+k^2} \tag{92-1}$$

Then, we apply equation (91) on (92), as follows

$$L\left\{\frac{1-\cos\ kt}{t}\right\} = \int_s^\infty \left(\frac{1}{x} - \frac{x}{x^2+k^2}\right)dx \tag{92-2}$$

$$= \left(\log x - \frac{1}{2}\log\left(x^2+k^2\right)\right)_s^\infty$$

$$= \left(\log\frac{x}{\sqrt{x^2+k^2}}\right)_s^\infty$$

$$= \lim_{x\to\infty} \log\frac{x}{\sqrt{x^2+k^2}} - \log\frac{s}{\sqrt{s^2+k^2}}$$

$$= \lim_{x\to\infty} \log\frac{1}{\sqrt{1+\frac{k^2}{x^2}}} - \log\frac{s}{\sqrt{s^2+k^2}}$$

$$= -\log\frac{s}{\sqrt{s^2+k^2}} = \frac{1}{2}\log\frac{s^2+k^2}{s^2} = \frac{1}{2}\log\left(1+\frac{k^2}{s^2}\right) \tag{92-3}$$

Example 52

Find the Laplace transform of

$$L\left\{\frac{e^{-at} - e^{-bt}}{t}\right\},$$ (93)

Solution

First,

$$L\left\{e^{-at} - e^{-bt}\right\} = \frac{1}{s+a} - \frac{1}{s+b}$$ (93-1)

Then

$$L\left\{\frac{e^{-at} - e^{-bt}}{t}\right\} = \int_s^\infty \left(\frac{1}{x+a} - \frac{1}{x+b}\right) dx$$ (93-2)

$$= \left[\log(x+a) - \log(x+b)\right]_s^\infty$$

$$= \left[\log\left(\frac{x+a}{x+b}\right)\right]_s^\infty = \lim_{x\to\infty}\log\left(\frac{x+a}{x+b}\right) - \log\left(\frac{s+a}{s+b}\right)$$

$$= -\log\left(\frac{s+a}{s+b}\right), \qquad s > -a, \qquad s > -b$$ (93-3)

Corollary

In equation (93-3), if a = 0 and b = 1 we get

$$L\left\{\frac{1-e^{-t}}{t}\right\} = \log\frac{s+1}{s} = \log\left(1+\frac{1}{s}\right), \qquad s > 0$$ (94)

Example 53

Prove that

$$\int_0^\infty \frac{F(t)}{t} dt = \int_0^\infty f(x) dx$$ (95)

Provided that the integrals converge.

Hence, deduce that

(i) $\quad \displaystyle\int_0^\infty \frac{\sin(t)}{t} dt = \frac{\pi}{2}$

(ii) $\quad \displaystyle\int_0^\infty \frac{e^{-t} - e^{-3t}}{t} dt = \log 3$

Solution

Since, from (91),

103

$$L\left\{\frac{F(t)}{t}\right\} = \int\limits_{s}^{\infty} f(x)dx$$

Then

$$L\left\{\frac{F(t)}{t}\right\} = \int\limits_{0}^{\infty} e^{-st}\frac{F(t)}{t}dt = \int\limits_{s}^{\infty} f(x)dx$$

Let s → 0 +, assuming that the integrals converge, then

$$\int\limits_{0}^{\infty} e^{-st}\frac{F(t)}{t}dt = \int\limits_{0}^{\infty} f(x)dx$$

(i) Take F(t) = sin t.

Then

$$f(s) = \frac{1}{s^2 + 1}$$

We get

$$\int\limits_{0}^{\infty} e^{-st}\frac{\sin(t)}{t}dt = \int\limits_{0}^{\infty} \frac{dx}{s^2 + 1}$$

$$= \left[\ \tan^{-1}x\ \right]_{0}^{\infty} = \frac{\pi}{2} \tag{95-1}$$

(ii) Take F(t) = $e^{-t} - e^{-3t}$.

Then

$$f(s) = \frac{1}{s+1} - \frac{1}{s+3}$$

We get

$$\int\limits_{0}^{\infty} e^{-st}\left(\frac{e^{-t} - e^{-3t}}{t}\right)dt = \int\limits_{s}^{\infty} \left(\frac{1}{x+1} - \frac{1}{x+3}\right)dx$$

$$= \left[\ \log\frac{x+1}{x+3}\ \right]_{s}^{\infty}$$

$$= -\log\frac{x+1}{x+3} == \log\frac{x+3}{x+1} \tag{95-2}$$

Taking the limit as s → 0 we get

$$\int\limits_{0}^{\infty} e^{-st}\left(\frac{e^{-t} - e^{-3t}}{t}\right)dt = \lim_{x \to 0} \log\frac{x+3}{x+1} = \log 3$$

2.12. Transforms of periodic functions

If $F(t)$ is a **periodic** function of period **a**, i.e., $F(t + a) = F(t)$ and if $F(t)$ is **sectionally continuous** over a period, then

$$f(s) = \frac{\int_0^s e^{-st}F(t)dt}{1 - e^{-as}} \qquad (96)$$

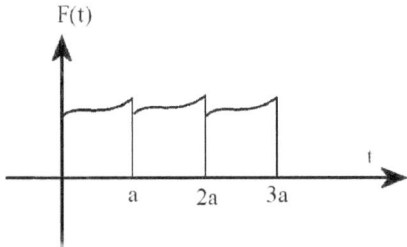

Proof

$$\int_0^\infty e^{-st}F(t)dt = \int_0^a e^{-st}F(t)dt + \int_a^{2a} e^{-st}F(t)dt + ... \qquad (96\text{-}1)$$

$$= \sum_{n=0}^\infty \int_{na}^{(n+1)a} e^{-st}F(t)dt \qquad (96\text{-}2)$$

Put $\qquad\qquad\qquad t = \tau + na$

Then, $\qquad\qquad\quad F(\tau + na) = F(\tau)$

Thus, equation (96-2) becomes

$$f(s) = \sum_{n=0}^\infty \int_0^a e^{-s(\tau+na)} \; F(\tau + na) \; d\tau \qquad (96\text{-}3)$$

Using the shift property, equation (96-3), can be written as

$$f(s) = \sum_{n=0}^\infty e^{-san} \int_0^a e^{-st} \; F(t) \; dt$$

$$= (1 + e^{-sa} + e^{-2sa} + e^{-3sa} + ..) \int_0^a e^{-st} \; F(t) \; dt$$

$$= (1 - e^{-sa})^{-1} \int_0^a e^{-st} \; F(t) \; dt == \dfrac{\displaystyle\int_0^a e^{-st} \; F(t) \; dt}{1 - e^{-sa}} \qquad (96\text{-}4)$$

Example 54

Find the Laplace transform of the square wave or the **Meander function** $M(c,t)$ defined by

$$
\begin{array}{llll}
M(c,t) & = \; 1 & & 0 < t < c \\
 & = -1 & & c < t < 2c
\end{array} \qquad (97)
$$

And that
$$M(c, t + 2c) = M(c,t)$$

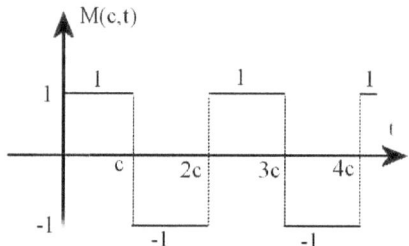

Since the period of the function is 2c, then

$$L\{M(c,t)\} = \dfrac{\displaystyle\int_0^{2c} e^{-st} M(c,t) dt}{1 - e^{-2cs}} \qquad (97\text{-}1)$$

Substituting by the values of $M(c,t)$ from (97) in (97-1), we get

$$\int_0^{2c} e^{-st} M(c,t) dt = \int_0^c e^{-st} dt - \int_c^{2c} e^{-st} dt$$

$$= \left[\dfrac{e^{-st}}{-s} \right]_0^c + \left[\dfrac{e^{-st}}{s} \right]_c^{2c} = \dfrac{1}{s} \left(1 - e^{-cs}\right)^2 \qquad (97\text{-}2)$$

106

From (97-1) and (97-2), we get

$$L\{M(c,t)\} = \frac{\int_{0}^{2c} e^{-st} M(c,t)dt}{1-e^{-2cs}} = \frac{1}{s}.\frac{\left(1-e^{-cs}\right)^{2}}{1-e^{-2cs}}$$

$$= \frac{1}{s}.\frac{\left(1-e^{-cs}\right)^{2}}{\left(1-e^{-cs}\right)\left(1+e^{-cs}\right)} = \frac{1}{s}.\frac{1-e^{-cs}}{1+e^{-cs}} \qquad (97\text{-}3)$$

$$= \frac{1}{s}.\frac{e^{\frac{cs}{2}}-e^{-\frac{cs}{2}}}{e^{\frac{cs}{2}}+e^{-\frac{cs}{2}}} = \frac{1}{s}.\tanh\left(\frac{cs}{2}\right),\ s>0 \qquad (97\text{-}4)$$

$$L\{M(c,t)\} = \frac{1}{s}.\tanh\left(\frac{cs}{2}\right) \qquad (97\text{-}5)$$

Example 55

Find the Laplace transform of the triangular wave or the function H(c,t) defined as follows :

$$
\begin{aligned}
H(c,t) \quad &= t & 0 < t < c \\
&= 2c - t & c < t < 2c
\end{aligned} \qquad (98)
$$

And that

$$H(c, t+2c) = H(c,t)$$

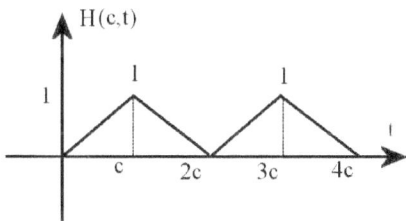

It is evident that H(c,t) at any t is the **integral of the function** M(c,t) from 0 to t.
Thus, by dividing (97 5) by s, which corresponds to integration of the object function, we get

$$L\{H(c,t)\} = \frac{1}{s^{2}}.\tanh\left(\frac{cs}{2}\right) \qquad (98\text{-}1)$$

Example 56

Find the Laplace transform of the periodic function F(t) defined by

107

$$F(t) \quad = t \qquad 0 < t < 1 \tag{99}$$

And that

$$F(t + 1) = F(t)$$

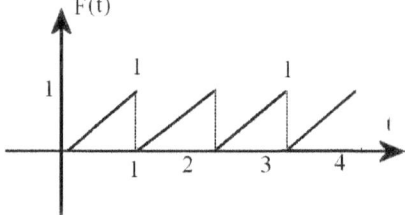

Solution

From equations (96) and (99), we get

$$f(s) = \frac{\int_0^1 t\ e^{-st}dt}{1 - e^{-s}} \tag{99-1}$$

First, we will deal with the denominator of (99-1) as follows

$$\int_0^1 t\ e^{-st}dt = \int_0^1 t\ \frac{d\left(e^{-st}\right)}{-s} \tag{99-2}$$

Integrating (99-2) by parts, we get

$$\int_0^1 t\ e^{-st}dt = \left[\frac{te^{-st}}{-s}\right]_0^1 - \int_0^1 \frac{e^{-st}}{-s}dt$$

$$= \left[\frac{te^{-st}}{-s}\right]_0^1 - \left[\frac{e^{-st}}{s^2}\right]_0^1$$

$$= \frac{e^{-s}}{-s} - \frac{e^{-s} - 1}{s^2} \tag{99-3}$$

Second, substituting by equation (99-3) in (99-1), we get

$$f(s) = \frac{1}{1-e^{-s}}\left(\frac{e^{-s}}{-s} - \frac{e^{-s}-1}{s^2}\right) = \frac{1}{s^2} - \frac{e^{-s}}{s(1-e^{-s})} \tag{99-4}$$

Example 57

Find the Laplace transform of the intermittent sine wave defined by

$$
\begin{aligned}
F(t) \quad &= \sin t & 0 < t < \pi \\
&= 0 & \pi < t < 2\pi
\end{aligned}
\tag{100}
$$

And that

$$F(t + 2\pi) = F(t)$$

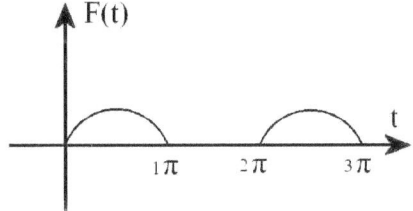

Solution

From equations (96) and (100), we get

$$f(s) = \frac{\displaystyle\int_0^{2\pi} F(t)\ e^{-st}dt}{1-e^{-2\pi s}} = \frac{\displaystyle\int_0^{2\pi} e^{-st}\sin(t)dt}{1-e^{-2\pi s}} \tag{100-1}$$

First, we start with the integral in the denominator of (100-1)

$$\int_0^{\pi} e^{-st}\sin(t)dt = \frac{1+e^{-\pi s}}{s^2+1} \tag{100-2}$$

Second, substituting from (100-2) into (100-1), we get

$$f(s) = \frac{1}{1-e^{-2\pi s}}\left[\frac{1+e^{-\pi s}}{s^2+1}\right]$$

$$= \frac{1}{\left(1-e^{-\pi s}\right)\left(1+e^{-\pi s}\right)}\left[\frac{1+e^{-\pi s}}{s^2+1}\right] = \frac{1}{\left(1-e^{-\pi s}\right)\left(s^2+1\right)} \tag{100-3}$$

Example 58

Find the Laplace transform of the full-wave rectification $|\sin t|$ of the sine function.

109

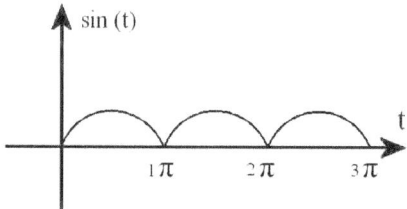

Solution

It is evident that in the full wave rectification the ordinates are the **sum** of the ordinates of F(t) of Example 57 in addition to ordinates of F(t) displaced to the right a distance π.

Hence, from equation (100-3) we get

$$L\left\{\left|\sin t\right|\right\} = \frac{1}{\left(1 - e^{-\pi s}\right)\left(s^2 + 1\right)} + \frac{e^{-\pi s}}{\left(1 - e^{-\pi s}\right)\left(s^2 + 1\right)} \tag{101}$$

$$= \frac{1 + e^{-\pi s}}{\left(1 - e^{-\pi s}\right)\left(s^2 + 1\right)}$$

$$= \frac{1}{s^2 + 1}\left\{\frac{e^{\frac{\pi s}{2}} + e^{-\frac{\pi s}{2}}}{e^{\frac{\pi s}{2}} - e^{-\frac{\pi s}{2}}}\right\} = \frac{1}{s^2 + 1}\coth\left(\frac{\pi s}{2}\right) \tag{101-1}$$

2.13. The product theorem—Convolution

This theorem aims at finding the inverse transform of the product of two transforms.

Let $F_1(t)$ and $F_2(t)$ be two sectionally **continuous** functions in every finite interval $0 \le t \le T$ and of the order of e^{at} as $t \to \infty$.

Take $s > a$ and let

$$L\left\{F_1(t)\right\} = f_1(s)$$
$$L\left\{F_2(t)\right\} = f_2(s) \tag{102}$$

Then

$$L^{-1}\left\{f_1(s)\ f_2(s)\right\} = F_1(t) * F_2(t) \tag{102-1}$$

Where

$$F_1(t) * F_2(t) = \int_0^t F_1(t - \lambda)\ F_2(\lambda)\ d\lambda \tag{102-2}$$

$F_1(t)* F_2(t)$ as defined above is called the **convolution** of the two functions and the last integral is known as a **convolution integral** or Faltung integral (which means folding in German).

Proof

By definition, the two Laplace transforms of $F_1(t)$ and $F_2(t)$ are give as

$$f_1(s) = \int_0^\infty e^{-sx} F_1(x) \ dx$$

$$f_2(s) = \int_0^\infty e^{-sy} F_2(y) \ dy \qquad (102\text{-}3)$$

From (102-3), the product of f_1 (s) and f_2 (s) is

$$f_1(s) \ f_2(s) = \int_0^\infty e^{-sx} F_1(x) \ dx \int_0^\infty e^{-sy} F_2(y) \ dy \qquad (102\text{-}4)$$

$$= \int_0^\infty \int_0^\infty e^{-s(x+y)} F_1(x) \ F_2(y) \ dxdy \qquad (102\text{-}5)$$

The integration in (102-5) is *being* carried over the positive quadrant $x > 0$, $y > 0$.

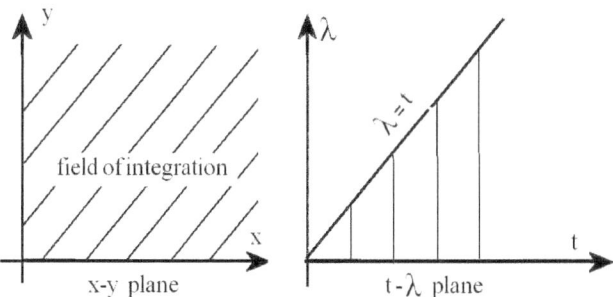

x-y plane   t-λ plane

Now, let us make the following transformation of coordinates

$$x = t - \lambda \qquad \text{i.e.,} \qquad t = x + \lambda$$
$$y = \lambda \qquad \text{i.e.,} \qquad \lambda = y \qquad (102\text{-}6)$$

Since $x \geq 0$, then $t - \lambda \geq 0$, i. e., $t \geq \lambda$.

Hence, the field of integration in the x-y plane transforms in the $t - \lambda$ plane into the field between the positive t-axis and the straight line $\lambda = t$.

111

Now the element of area dx dy transforms into

$$\text{element of area} = dx\, dy = \begin{vmatrix} \dfrac{\partial x}{\partial t} & \dfrac{\partial y}{\partial t} \\ \dfrac{\partial x}{\partial \lambda} & \dfrac{\partial y}{\partial \lambda} \end{vmatrix} dt d\lambda \qquad (102\text{-}7)$$

From (102-6), by partial differentiation, we get

$$\partial x/\partial t = 1$$
$$\partial x/\partial \lambda = -1$$
$$\partial y/\partial t = 0$$
$$\partial y/\partial \lambda = 1 \qquad (102\text{-}7a)$$

Thus, the element of area dx.dy, in equation (102-7) becomes

$$\text{element of area} = dx\, dy = \begin{vmatrix} 1 & 0 \\ -1 & 1 \end{vmatrix} dt d\lambda = dt d\lambda \qquad (102\text{-}7b)$$

Hence, the double integral (102-5) transforms into

$$f_1(s)\ f_2(s) = \int_0^\infty \int_0^\infty e^{-s(x+y)} F_1(x)\ F_2(y)\ dx dy$$

$$= \int_0^\infty \int_0^t e^{-st} F_1(t-\lambda)\ F_2(\lambda)\ dt d\lambda$$

$$= \int_0^\infty e^{-st} \left[\int_0^t F_1(t-\lambda)\ F_2(\lambda) d\lambda \right] dt$$

$$= L\left\{ \int_0^t F_1(t-\lambda)\ F_2(\lambda) d\lambda \right\} = L\{F_1 * F_2\} = L\{F_2 * F_1\} \qquad (102\text{-}8)$$

The last portion in (102-8) is obtained by symmetry of the integration process.

Hence, we have the following result:

$$L^{-1}\{ f_1(s) \} = F_1(t)$$
$$L^{-1}\{ f_2(s) \} = F_2(t)$$

Then

$$L^{-1}\{ f_1(s)\ f_2(s) \} = F_1(t) * F_2(t)$$

$$= \int_0^t F_1(t-\lambda)\ F_2(\lambda)\ d\lambda = \int_0^t F_2(t-\lambda)\ F_1(\lambda)\ d\lambda \qquad (102\text{-}9)$$

This formula is known in **operational calculus** as the **Borel** formula.

Example 59

Find the Inverse Laplace transform of

$$L^{-1}\left\{\frac{1}{s^2-a^2}\right\}$$ (103)

Solution

From equation (102-1), we write

$$L^{-1}\left\{\frac{1}{s^2-a^2}\right\} = L^{-1}\left\{\frac{1}{s-a}\cdot\frac{1}{s+a}\right\} = e^{at} * e^{-at}$$

And from equation (102-2), we write

$$e^{at} * e^{-at} = \int_0^t e^{a(t-\lambda)}e^{-a\lambda}d\lambda = e^{at}\int_0^t e^{-2a\lambda}d\lambda$$

$$= \frac{e^{at}}{-2a}\left[e^{-2a\lambda}\right]_0^t$$

$$= -\frac{e^{at}}{2a}\left[e^{-2at}-1\right] = \frac{e^{at}e^{-at}}{2a}\left[e^{at}-e^{-at}\right] = \frac{1}{a}\left[\frac{e^{at}-e^{-at}}{2}\right]$$

$$= \frac{1}{a}\sinh(at)$$ (103-1)

Example 60

Find the Inverse Laplace transform of

$$L^{-1}\left\{\frac{a}{\left(s^2+a^2\right)^2}\right\}$$ (104)

Solution

From equation (102-1), we write

$$L^{-1}\left\{\frac{s}{\left(s^2+a^2\right)^2}\right\} = L^{-1}\left\{\frac{s}{s^2+a^2}\cdot\frac{1}{s^2+a^2}\right\} = \cos(at) * \frac{1}{a}\sin(at)$$ (104-1)

And from equation (102-2), we write

$$\cos(at)*\frac{1}{a}\sin(at)=\frac{1}{a}\int_0^t \cos\ a(t-\lambda)\frac{1}{a}\sin\ (a\lambda)\ d\lambda \qquad (104\text{-}2)$$

Since, the algebraic properties of the sine function provide for the following expansions

$$\begin{array}{llll}
\sin at & = \sin a(t\text{-}\lambda + \lambda) & = \sin a(t\text{-}\lambda)\cos a\lambda + \cos a(t\text{-}\lambda)\sin a\lambda & (104\text{-}3)\\
\sin a(t\text{-}2\lambda) & = \sin a(t\text{-}\lambda - \lambda) & = \sin a(t\text{-}\lambda)\cos a\lambda - \cos a(t\text{-}\lambda)\sin a\lambda & (104\text{-}4)
\end{array}$$

then, the subtraction of (104-4) from (104-3) gives

$$\sin at - \sin a(t\text{-}2\lambda) = 2\cos a(t\text{-}\lambda)\sin a\lambda \qquad (104\text{-}5)$$

From (104-5) and (104-2), we get

$$\cos(at)*\frac{1}{a}\sin(at)=\frac{1}{2a}\int_0^t\left[\sin\ at-\sin a(t-2\lambda)\right]\ d\lambda$$

$$=\frac{1}{2a}\left[\lambda\sin\ at+\frac{1}{-2a}\cos a(t-2\lambda)\right]_{\lambda=0}^{\lambda=t}$$

$$=\frac{1}{2a}\left[t\sin\ at-0+\frac{1}{-2a}(\cos a(t-2t)-\cos a(t))\right]=\frac{1}{2a}t\sin at$$

$$L^{-1}\left\{\frac{s}{\left(s^2+a^2\right)^2}\right\}==\frac{1}{2a}t\sin at \qquad (104\text{-}6)$$

2.14. Application of the product theorem to the solution of differential and integral equations

Example 61

Solve the differential equation

$$Y''(t)+k^2\ Y(t)=F(t) \qquad (105)$$

where k is a constant.

Solution

The Laplace Transform of (105) is

$$s^2\ y(s)-s\ Y(0)-Y'(0)+k^2\ y(s)=f(s) \qquad (105\text{-}1)$$

Put

$$Y(0)=A$$
$$Y'(0)=B$$

114

Thus, equation (105-1) is simplified to

$$(s^2 + k^2) \, y(s) + A \, s + B + f(s) \tag{105-2}$$

Therefore, the inverse Laplace transform of (105-2) is

$$y(s) = \frac{As}{s^2 + k^2} + \frac{B}{s^2 + k^2} + \frac{f(s)}{s^2 + k^2} \tag{105-3}$$

The last term in equation (105-3) is the product of two functions and will be dealt with by equation (102-2). Thus the inverse Laplace transform of (105-3) is

$$\begin{aligned}
Y(t) = L^{-1} &\left\{ \frac{As}{s^2 + k^2} + \frac{B}{s^2 + k^2} + \frac{f(s)}{s^2 + k^2} \right\} \\
&= A \cos kt + \frac{B}{k} \sin kt + \frac{1}{k} \sin kt * F(t) \\
&= A \cos kt + \frac{B}{k} \sin kt + \frac{1}{k} \int_0^t \sin k(t - \lambda) F(\lambda) d\lambda
\end{aligned} \tag{105-4}$$

This method is particularly useful in the case when the transform of $F(t)$ is difficult to obtain, for example, as in the case of an **impressed voltage given by an oscillogram**.

The convolution integral can, however, be calculated **numerically**.

Example 62

Solve the integral equation

$$Y(t) = 4 \, \sin t + \int_0^t \sin(t - \lambda) Y(\lambda) d\lambda \tag{106}$$

By an integral equation we mean an equation in which the unknown function lies under the integral sign.

Solution

Here, the integral is of the **convolution type**. Integrals of the convolution type appear in many practical problems.

Equation (106) can he written in the form

$$Y(t) = 4 \sin t + \sin t * Y(t) \tag{106-1}$$

Transforming we get

$$y(s) = \frac{4}{s^2 + 1} + \frac{1}{s^2 + 1} y(s)$$

$$y(s)\left(1 - \frac{1}{s^2 + 1}\right) = \frac{4}{s^2 + 1}$$

i.e.,
$$y(s) = \frac{4}{s^2}$$
(106-2)

The inverse of (106-2)

$$Y(t) = L^{-1}\left\{\frac{4}{s^2}\right\} = 4t$$
(106-3)

Example 63

Solve the integral equation

$$Y(t) = \frac{1}{2} \ t^2 - \int_0^t (t - \lambda)Y(\lambda)d\lambda$$
(107)

Solution

With the help of equation (102-1) for convolution integrals, equation (107) is reduced to

$$Y(t) = \frac{1}{2} \ t^2 - t * Y(t)$$
(107-1)

The Laplace transform of (107-1) is thus

$$y(s) = \frac{1}{s^3} - \frac{1}{s^2} y(s)$$
(107-2)

$$y(s) \left[1 + \frac{1}{s^2}\right] = \frac{1}{s^3}$$

$$y(s) = \frac{1}{s^3} \frac{s^2}{s^2 + 1} = \frac{1}{s(s^2 + 1)} = \frac{1}{s} - \frac{s}{s^2 + 1}$$
(107-3)

The inverse Laplace Transform of (017-3) is

$$Y(t) = L^{-1}\left(\frac{1}{s} - \frac{s}{s^2 + 1}\right) = 1 - \cos t$$
(107-4)

Example 64

Solve the integro- differential equation

116

$$Y'(t) = t + \int_0^t Y(t-\lambda) \, \cos\lambda \, d\lambda \tag{108}$$

Given that $Y(0) = 6$.

Solution

With the help of the theory developed in (102), equation (108) can be written as

$$Y'(t) = t + Y(t) * \cos t \tag{108-1}$$

With the Laplace transform

$$s \ y(s) - Y(0) = \frac{1}{s^2} + y(s).\frac{s}{s^2 + 1} \tag{108-2}$$

Put $Y(0) = 6$ and arrange the terms in (108-2), we get

$$y(s) \left[s - \frac{s}{s^2 + 1} \right] = 6 + \frac{1}{s^2}$$

$$y(s) = \left(\frac{s^2 + 1}{s^3} \right) \left(6 + \frac{1}{s^2} \right) = \frac{6}{s} + \frac{7}{s^3} + \frac{1}{s^5} \tag{108-3}$$

The inverse Laplace transform of (108-3) is

$$Y(t) = L^{-1} \left(\frac{6}{s} + \frac{7}{s^3} + \frac{1}{s^5} \right) = 6 + \frac{7}{2} t^2 + \frac{1}{24} t^4 \tag{108-4}$$

2.15. Power series method for the determination of transforms and inverse transforms

In many problems, expansion into **infinite series** is helpful in finding Laplace transforms and inverse transforms. The following are additional examples on the same principle. A list of well-known expansions is given here for reference.

(1) $\dfrac{1}{1-x} = \sum\limits_{n=0}^{\infty} x^n$ $\qquad\qquad |x| < 1$ $\qquad\qquad$ (108-4.1)

(2) $\dfrac{1}{1+x} = \sum\limits_{n=0}^{\infty} (-1)^n x^n$ $\qquad\qquad |x| < 1$ $\qquad\qquad$ (108-4.2)

(3) $e^x = \sum\limits_{n=0}^{\infty} \dfrac{x^n}{n!}$ $\qquad\qquad$ all values of x $\qquad\qquad$ (108-4.3)

(4) $\cos x = \sum\limits_{n=0}^{\infty} \dfrac{(-1)^n x^{2n}}{(2n)!}$ $\qquad\qquad$ all values of x $\qquad\qquad$ (108-4.4)

(5) $\sin x = \sum_{n=0}^{\infty} \frac{(-1)^n x^{2n+1}}{(2n+1)\,!}$ all values of x (108-4.5)

(6) $\cosh x = \sum_{n=0}^{\infty} \frac{x^{2n}}{(2n)\,!}$ all values of x (108-4.6)

(7) $\sinh x = \sum_{n=0}^{\infty} \frac{x^{2n+1}}{(2n+1)\,!}$ all values of x (108-4.7)

(8) $\tan^{-1} x = \sum_{n=0}^{\infty} \frac{(-1)^n x^{2n+1}}{2n+1}$ $|x| \le 1$ (108-4.8)

(9) $\log(1+x) = \sum_{n=0}^{\infty} \frac{(-1)^n x^{n+1}}{n+1}$ $-1 < x \le 1$ (108-4.9)

(10) $\log\dfrac{1+x}{1-x} = 2\sum_{n=0}^{\infty} \frac{x^{2n+1}}{2n+1}$ $|x| < 1$ (108-4.10)

Example 65

Evaluate $\qquad\qquad L^{-1}\left\{\dfrac{1}{s^3 \cosh 2s}\right\}$ (109)

Then, compute F(12)

Solution

Let us expand the cosh 2s into its exponential components as follows

$$\frac{1}{s^3 \cosh 2s} = \frac{2}{s^3\left(e^{2s}+e^{-2s}\right)} = \frac{2e^{-2s}}{s^3\left(1+e^{-4s}\right)}$$

$$= \sum_{n=0}^{\infty} \frac{2}{s^3} e^{-2s}(-1)^n e^{-4ns} = \sum_{n=0}^{\infty} \frac{2}{s^3}(-1)^n e^{-(4n+2)s} \qquad (109\text{-}1)$$

The inverse Laplace transform of (109-1) is

$$L^{-1}\left\{\frac{1}{s^3 \cosh 2s}\right\} = L^{-1}\left\{\sum_{n=0}^{\infty} \frac{2}{s^3}(-1)^n e^{-(4n+2)s}\right\} \qquad (109\text{-}2)$$

We have shown in equation (57-3) that

$$L^{-1}\left\{e^{-st} f(s)\right\} = U(t-a)\,F(t-a), \quad s > 0 \qquad (109\text{-}2a)$$

Therefore, equation (109-2) is shifted by the **unit step** function U(t - 4n – 2) where

$$f(s) = \frac{2}{s^3} \tag{109-2b}$$

$$F(t) = L^{-1}\left\{\frac{2}{s^3}\right\} = t^2 \tag{109-2c}$$

And the shifted object function is

$$F(t - a) = (t - 4n - 2)^2 \tag{109-2d}$$

Substituting from (109-2b), (109-2c), and (109-2d), in (109-2), we get

$$L^{-1}\left\{\frac{1}{s^3 \cosh 2s}\right\} = \sum_{n=0}^{\infty} (-1)^n (t - 4n - 2)^2 U(t - 4n - 2) \tag{109-3}$$

Upon expansion of the summation in (109-3), we get

$$F(t) = L^{-1}\left\{\frac{1}{s^3 \cosh 2s}\right\} = (t - 2)^2 U(t - 2) - (t - 6)^2 U(t - 6) + (t - 10)^2 U(t - 10) - .. \tag{109-4}$$

Notice that the series in the right hand side terminates once the **argument of the function U is negative**.

Hence for t =12, (109-4) gives

$$F(12) = (t - 2)^2 U(t - 2) - (t - 6)^2 U(t - 6) + (t - 10)^2 U(t - 10) - (t - 14)^2 U(12 - 14)$$
$$= 100 - 36 + 4 = 68$$

Example 66

Evaluate $\qquad\qquad L^{-1}\left\{\log\frac{s+1}{s-1}\right\}$ $\qquad\qquad\qquad$ (110)

Solution

We expand the log function in infinite series as follows

$$\log\frac{s+1}{s-1} = \log\frac{1 + \frac{1}{s}}{1 - \frac{1}{s}} \tag{110-1}$$

Thus, equation (110-1) takes the form of, $\log\dfrac{1+x}{1-x} = 2\sum\limits_{n=0}^{\infty}\dfrac{x^{2n+1}}{2n+1}$, and can be expanded as follows

$$\log\dfrac{1+\dfrac{1}{s}}{1-\dfrac{1}{s}} = 2\sum\limits_{n=0}^{\infty}\dfrac{\left(\dfrac{1}{s}\right)^{2n+1}}{(2n+1)} = 2\sum\limits_{n=0}^{\infty}\dfrac{1}{(2n+1)s^{2n+1}} \qquad (110\text{-}2)$$

We will now use equation (13) with gives

$$L\{t^{n}\} \ \ = \dfrac{\Gamma(n+1)}{s^{n+1}} = \dfrac{n!}{s^{n+1}}, \qquad s > 0 \qquad (110\text{-}3)$$

Hence,

$$L^{-1}\left\{\dfrac{1}{s^{2n+1}}\right\} = \dfrac{t^{2n}}{(2n)!} \qquad (110\text{-}4)$$

Therefore, from (110-3) and (110-4), equation (110-2) becomes

$$L^{-1}\left\{\log\dfrac{1+\dfrac{1}{s}}{1-\dfrac{1}{s}}\right\} = 2L^{-1}\left\{\sum\limits_{n=0}^{\infty}\dfrac{1}{(2n+1)s^{2n+1}}\right\} = 2\left\{\sum\limits_{n=0}^{\infty}\dfrac{t^{2n}}{(2n+1)(2n)!}\right\}$$

$$= \dfrac{2}{t}\left\{\sum\limits_{n=0}^{\infty}\dfrac{t^{2n+1}}{(2n+1)!}\right\} \qquad (110\text{-}5)$$

We have stated above that

$$\sinh x = \sum\limits_{n=0}^{\infty}\dfrac{x^{2n+1}}{(2n+1)!} \qquad \text{all values of x} \qquad (110\text{-}6)$$

Therefore, equation (110-5) becomes

$$L^{-1}\left\{\log\dfrac{1+\dfrac{1}{s}}{1-\dfrac{1}{s}}\right\} = \dfrac{2}{t}\sinh t \qquad (110\text{-}7)$$

Example 67

Find $\qquad L\left\{\displaystyle\int_{0}^{t}\dfrac{\sin u}{u}\,du\right\} \qquad (111)$

120

Solution

We expand the sine function according to the formula given above

$$\frac{\sin x}{x} = \frac{1}{x} \sum_{n=0}^{\infty} \frac{(-1)^n x^{2n+1}}{(2n+1)\ !} \tag{111-1}$$

$$= \frac{1}{x}\left(x - \frac{x^3}{3!} + \frac{x^5}{5!} - . \right) = 1 - \frac{x^2}{3!} + \frac{x^4}{5!} -$$

Therefore, the integral in (111) becomes

$$\int_0^t \frac{\sin u}{u} du = \int_0^t \left(1 - \frac{u^2}{3!} + \frac{u^4}{5!} - \right) du = t - \frac{t^3}{3.3!} + \frac{t^5}{5.5!} - \tag{111-2}$$

Thus,

$$L\left\{ \int_0^t \frac{\sin u}{u} du \right\} = L\left\{ t - \frac{t^3}{3.3!} + \frac{t^5}{5.5!} - \right\} = \frac{1}{s^2} - \frac{1}{3s^4} + \frac{1}{5s^6} - \tag{111-3}$$

$$= \frac{1}{s}\left(\frac{1}{s} - \frac{1}{3s^3} + \frac{1}{5s^5} - \right) \tag{111-4}$$

We the use the expansion stated above

$$\tan^{-1} x = \sum_{n=0}^{\infty} \frac{(-1)^n x^{2n+1}}{2n+1} \tag{111-5}$$

Hence, by equation (111-5), equation (111-4) becomes

$$L\left\{ \int_0^t \frac{\sin u}{u} du \right\} == \frac{1}{s}\left(\frac{1}{s} - \frac{1}{3s^3} + \frac{1}{5s^5} - \right) = \frac{1}{s} \tan^{-1} \frac{1}{s} \tag{111-6}$$

Example 68

If $s > 0$ and $n > 1$ prove that

$$L\left\{ \frac{t^{n-1}}{1 - e^{-t}} \right\} == \Gamma(n)\left(\frac{1}{s^n} + \frac{1}{(s+1)^n} + \frac{1}{(s+2)^n} ... \right) \tag{112}$$

Solution

The Binomial theorem allow us to write

$$\frac{t^{n-1}}{1-e^{-t}} = t^{n-1}\left(1-e^{-t}\right)^{-1} = t^{n-1}\left(1+e^{-t}+e^{-2t}+..\right) \tag{112-1}$$

The exponentials in (112-1) shift the Laplace transforms with the coefficients of t as follows

$$L\left\{\frac{t^{n-1}}{1-e^{-t}}\right\} = L\left\{t^{n-1}\left(1+e^{-t}+e^{-2t}+..\right)\right\} = \frac{\Gamma(n)}{s^n} + \frac{\Gamma(n)}{(s+1)^n} + \frac{\Gamma(n)}{(s+2)^n} +.. \tag{112-2}$$

$$= \Gamma(n)\left(\frac{1}{s^n} + \frac{1}{(s+1)^n} + \frac{1}{(s+2)^n} +..\right) \tag{112-3}$$

Example 69

Find $\qquad L^{-1}\left(\dfrac{e^{-1/s}}{s}\right)$ $\qquad\qquad$ (113)

Solution

The series expansion of (113) gives

$$L^{-1}\left(\frac{e^{-1/s}}{s}\right) = L^{-1}\left\{\frac{1}{s}\left(1-\frac{1}{s}+\frac{1}{2!s^2}-\frac{1}{3!s^3}+...\right)\right\}$$

$$= 1-t+\frac{t^2}{1^2.2^2}-\frac{t^3}{1^2.2^2.3^2.}+.. \tag{113-1}$$

We will arrange the terms in (113-1) such that we could substitute the Bessel Function for the series expansion. Therefore, (113-1) becomes

$$L^{-1}\left(\frac{e^{-1/s}}{s}\right) = 1-t+\frac{t^2}{1^2.2^2}-\frac{t^3}{1^2.2^2.3^2.}+.. = 1-\frac{\left(2t^{1/2}\right)^2}{2^2}+\frac{\left(2t^{1/2}\right)^4}{2^2.4^2}-\frac{\left(2t^{1/2}\right)^6}{2^2.4^2.6^2}+$$

$$= J_0\left(2\sqrt{t}\right) \tag{113-2}$$

2.16. The error function or probability integral

This **error function** or **probability integral** is a tabulated function denoted by "erf(x)" and is defined as follows

$$\mathrm{erf}(x) = \frac{2}{\sqrt{\pi}}\int_0^x e^{-u^2}\,du \tag{114}$$

$$= \frac{2}{\sqrt{\pi}} \; X \; \text{shaded area}$$

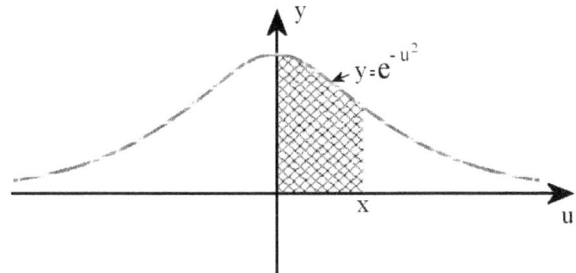

Expanding e^{-u^2} in equation (114) in ascending powers of u and integrating term by term we get.

$$erf(x) = \frac{2}{\sqrt{\pi}} \int_0^x \left(1 - u^2 + \frac{u^4}{2!} - \frac{u^6}{3!} + .. \right) du$$

$$= \frac{2}{\sqrt{\pi}} \left(x - \frac{x^3}{3} + \frac{x^5}{2!.5} - \frac{x^7}{3!.7} + .. \right) \qquad (114\text{-}1)$$

Now, since $\int_0^\infty e^{-u^2} du = \frac{\sqrt{\pi}}{2}$, it follows that **erf**(x) → 1 as x → ∞

The **complementary** error function **erfc** (x) is defined as follows

$$erfc(x) = 1 - erf(x) = \frac{2}{\sqrt{\pi}} \left[\int_0^\infty e^{-u^2} du - \int_0^x e^{-u^2} du \right]$$

$$= \frac{2}{\sqrt{\pi}} \int_x^\infty e^{-u^2} du \qquad (114\text{-}2)$$

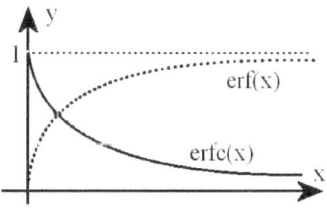

Example 70

Find $\quad L^{-1}\left\{\dfrac{1}{(s-1)\sqrt{s}}\right\}$ (115)

Solution

We have shown that

$$L\left\{t^{-\frac{1}{2}}\right\} = \sqrt{\dfrac{\pi}{s}} \quad \text{Hence,} \quad L^{-1}\left\{\dfrac{1}{\sqrt{s}}\right\} = \dfrac{1}{\sqrt{\pi t}}$$ (115-1)

First, we will use equation (115-1) along with the following previously proven rules:

(i) The **convolution** integral, equation (102-2), that gives $F_1(t)*F_2(t) = \displaystyle\int_0^t F_1(t-\lambda)\ F_2(\lambda)\ d\lambda$

(ii) Equation (8) that gives $L\{1\} = \dfrac{1}{s}$

(iii) Equation (14) that gives $L\{t^{-\frac{1}{2}}\} = \sqrt{\dfrac{\pi}{s}}, \quad s > 0$

(iv) Equation (28) that gives $L\left\{e^{at}\ F(t)\right\} = f(s-a)$

Thus, we can evaluate the following inverse Laplace transform

$$L^{-1}\left\{\dfrac{1}{(s-1)\sqrt{s}}\right\} = e^t * \dfrac{1}{\sqrt{\pi t}}$$ (115-2)

$$e^t * \dfrac{1}{\sqrt{\pi t}} = \dfrac{1}{\sqrt{\pi}}\int_0^t e^{t-\lambda}\dfrac{1}{\sqrt{\lambda}}d\lambda$$

$$= \dfrac{e^t}{\sqrt{\pi}}\int_0^t e^{-\lambda}\dfrac{1}{\sqrt{\lambda}}d\lambda$$ (115-3)

Put
$$\lambda = u^2$$
Then

$$\int_0^t e^{-\lambda}\dfrac{1}{\sqrt{\lambda}}d\lambda = \int_0^{\sqrt{t}}\dfrac{e^{-u^2}}{u}.2u\,du = 2\int_0^{\sqrt{t}} e^{-u^2}du$$ (115-4)

Therefore, equation (115-2) becomes

$$L^{-1}\left\{\dfrac{1}{(s-1)\sqrt{s}}\right\} = e^t.\dfrac{2}{\sqrt{\pi}}\int_0^{\sqrt{t}} e^{-u^2}du$$ (115-5)

$$= e^t.\mathrm{erf}\left(\sqrt{t}\right) \qquad\qquad (115\text{-}6)$$

Second, as we did above, we will use equation (115-1) to evaluate the following inverse Laplace transform

$$L^{-1}\left\{\frac{1}{s\sqrt{s}+1}\right\} == 1*\frac{e^{-t}}{\sqrt{\pi t}} \qquad\qquad (115\text{-}7)$$

As we stated earlier, equation (8) gives $L\{1\} = \dfrac{1}{s}$, equation (14) gives

$$L\{t^{-\frac{1}{2}}\} = \sqrt{\frac{\pi}{s}}, \quad s > 0 \quad, \text{ and equation (28) gives } L\left\{e^{at}\ F(t)\right\} = f(s-a). \text{ Therefore,}$$

$$1*\frac{e^{-t}}{\sqrt{\pi t}} = \frac{e^{-t}}{\sqrt{\pi}}\int_0^t \frac{e^{-\lambda}}{\sqrt{\lambda}}d\lambda \qquad\qquad (115\text{-}8)$$

We could have obtained (115-8) by changing s into s +1 in (115-6). Therefore, equation (115-7) becomes

$$L^{-1}\left\{\frac{1}{s\sqrt{s}+1}\right\} == e^{-t}e^t.\mathrm{erf}\left(\sqrt{t}\right) = \mathrm{erf}\left(\sqrt{t}\right) \qquad\qquad (115\text{-}9)$$

In sum:

(i) $\qquad L^{-1}\left\{\dfrac{1}{(s-1)\sqrt{s}}\right\} = e^t.\mathrm{erf}\left(\sqrt{t}\right)$

(ii) $\qquad L^{-1}\left\{\dfrac{1}{s\sqrt{s}+1}\right\} = \mathrm{erf}\left(\sqrt{t}\right)$

Example 71

Find $\qquad L\left\{e^{2t}\mathrm{erf}\left(\sqrt{t}\right)\right\}$ $\qquad\qquad (116)$

Solution

From (115-9), since $\qquad L\left\{\mathrm{erf}\left(\sqrt{t}\right)\right\} = \dfrac{1}{s\sqrt{s}+1}$ $\qquad\qquad (116\text{-}1)$

Shifting the object function by e^{2t} results in shifting s by -2 as follows

$$L\left\{e^{2t}\mathrm{erf}\left(\sqrt{t}\right)\right\} = \frac{1}{(s-2)\sqrt{s+1}-2} = \frac{1}{(s-2)\sqrt{s-1}} \qquad\qquad (116\text{-}2)$$

Example 72

Find

$$L\left\{t \ \text{erf}\left(2 \ \sqrt{t}\right)\right\}$$

(117)

Solution

From (115-9), since

$$L\left\{\text{erf}\left(\sqrt{t}\right)\right\} = \frac{1}{s\sqrt{s+1}}$$

(117-1)

Then

$$L\left\{\text{erf}\left(2 \ \sqrt{t}\right)\right\} = L\left\{\text{erf}\left(\sqrt{2t}\right)\right\}$$

$$= \frac{1}{4} \cdot \frac{1}{s\sqrt{\frac{s}{4}+1}} = \frac{2}{s\sqrt{s+4}}$$

(117-2)

We then use the differentiation of the Laplace transform in equation (79-1) which implies

$$\frac{\partial}{\partial s} L\left\{F(t)\right\} = \int_0^\infty \frac{\partial}{\partial s}\left\{e^{-st}F(t)\right\}dt = \frac{\partial}{\partial s} f(s) = L\left\{-t\,F(t)\right\} = f'(s)$$

(117-3)

Therefore, differentiating the Right Hand Side of equation (117-2) with respect to s leads to

$$L\left\{t \ \text{erf}\left(2 \ \sqrt{t}\right)\right\} = -\frac{d}{ds}\left(\frac{2}{s\sqrt{s+4}}\right) = -\frac{d}{ds}\left(2\left(s^3 + 4s^2\right)^{\frac{-1}{2}}\right)$$

(117-4)

$$= \left(-2\left(\frac{-1}{2}\right)\left(s^3 + 4s^2\right)^{\frac{-3}{2}}\left(3s^2 + 8s\right)\right)$$

$$= \frac{s(3s+8)}{\left(s^3 + 4s^2\right)^{\frac{3}{2}}} = \frac{3s+8}{s^2\left(s+4\right)^{\frac{3}{2}}}$$

(117-5)

Example 73

Find

$$L\left\{\text{erfc}\left(\sqrt{t}\right)\right\}$$

(118)

Solution

Since

$$\text{erfc}\left(\sqrt{t}\right) = 1 - \text{erfc}\left(\sqrt{t}\right)$$

Then

$$L\left\{ \operatorname{erfc}\left(\sqrt{t}\right)\right\} = L\left\{ 1 - \operatorname{erf}\left(\sqrt{t}\right)\right\} = \frac{1}{s} - \frac{1}{s\sqrt{s+1}}$$

$$= \frac{\sqrt{s+1}-1}{s\sqrt{s+1}} = \frac{\sqrt{s+1}-1}{s\sqrt{s+1}}\cdot\left(\frac{\sqrt{s+1}+1}{\sqrt{s+1}+1}\right)$$

$$= \frac{1}{\sqrt{s+1}\left(\sqrt{s+1}+1\right)} \tag{118-1}$$

Example 74

Find

$$L^{-1}\left\{\frac{1}{\sqrt{s+1}+1}\right\} \tag{119}$$

Solution

We write the fraction in (119) into its partial fractions form as follows

$$\frac{1}{\sqrt{s+1}+1} = \frac{1}{\sqrt{s+1}+1}\left(\frac{\sqrt{s+1}-1}{\sqrt{s+1}-1}\right) = \frac{\sqrt{s+1}-1}{s}$$

$$= -\frac{1}{s} + \frac{\sqrt{s+1}}{s} = -\frac{1}{s} + \frac{\sqrt{s+1}}{s}\left(\frac{\sqrt{s+1}}{\sqrt{s+1}}\right)$$

$$= -\frac{1}{s} + \frac{s+1}{s\sqrt{s+1}} = -\frac{1}{s} + \frac{1}{s\sqrt{s+1}} + \frac{1}{\sqrt{s+1}} \tag{119-1}$$

Therefore,

$$L^{-1}\left\{-\frac{1}{s} + \frac{1}{s\sqrt{s+1}} + \frac{1}{\sqrt{s+1}}\right\} = -1 + \operatorname{erf}\left(\sqrt{t}\right) + \frac{e^{-t}}{\sqrt{\pi t}}$$

$$= \frac{e^{-t}}{\sqrt{\pi t}} - \operatorname{erfc}\left(\sqrt{t}\right) \tag{119-2}$$

Example 75

Find

$$L\left\{\int_0^t \operatorname{erf}\left(\sqrt{u}\right)du\right\} \tag{120}$$

Solution

Since the integration of the object function between the limits 0 and t corresponds to the **division** of the transform by s, then, from equation (118-1), we get

$$L\left\{ \operatorname{erfc}\left(\sqrt{t}\right)\right\} = \frac{1}{\sqrt{s+1}\left(\sqrt{s+1}+1\right)}$$

Therefore

$$L\left\{ \int_0^t \operatorname{erfc}\left(\sqrt{t}\right)dt\right\} = \frac{1}{s}\cdot\frac{1}{\sqrt{s+1}\left(\sqrt{s+1}+1\right)}$$ (120-1)

2.17. The sine-integral function Si(t)

This is a tabulated function defined by

$$Si(t) = \int_0^t \frac{\sin u}{u}.du$$ (121)

It is represented by the shaded area under the curve $y = \dfrac{\sin u}{u}$ between 0 and t.

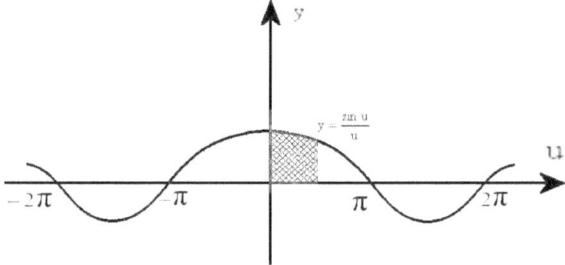

The function $y = \dfrac{\sin u}{u}$ is an even function which $\to 0$ as $u \to \infty$.

It intersects the u – axis at $u = \pm\,\pi, \pm\,2\pi, \pm\,3\pi,...$

Since $$L\left\{\sin t\right\} = \frac{1}{s^2 + 1}$$ (122)

Therefore, the division of the object function (sin t) by t implies the integration of the Right Hand Side of (122), as follows

$$L\left\{\frac{\sin t}{t}\right\} = \int_s^\infty \frac{dx}{x^2 + 1} = \left[\tan^{-1} x\right]_s^\infty$$ (122-1)

128

$$= \frac{\pi}{2} - \tan^{-1} s = \cot^{-1} s \qquad (122\text{-}2)$$

Again, applying the integration on the object function in (122-1) implies the division of the Right Hand Side of (122-2) by s, as follows

$$L\{Si(t)\} = L\left\{\int_0^t \frac{\sin u}{u}\,du\right\} = \frac{1}{s}\cot^{-1} s, \qquad s > 0 \qquad (122\text{-}3)$$

The following is an **alternative method** for finding $L\{Si(t)\}$

Put
$$F(t) = \int_0^t \frac{\sin u}{u}\,.du \qquad (122\text{-}3)$$

Then
$$\frac{d}{dt}F(t) = \frac{\sin t}{t}$$

Therefore
$$L\left\{t\frac{d}{dt}F(t)\right\} = L\{\sin t\}$$

$$L\left\{t\frac{d}{dt}F(t)\right\} = \frac{1}{s^2+1} = -\frac{d}{ds}\left[s\ f(s) - F(0)\right]$$

Therefore
$$\frac{d}{ds}\left[s\ f(s)\right] = -\frac{1}{s^2+1}$$

Integrating, we get
$$s\ f(s) = -\tan^{-1} s + C \qquad (122\text{-}4)$$

According to the initial value theorem
$$\lim_{s\to\infty}\ s\,f(s) - F(0) = \lim_{t\to 0}\ F(t)$$

And, since
$$\lim_{t\to 0}\ F(t) = F(0) = 0$$

Then
$$\lim_{s\to\infty}\ s\,f(s) = 0$$

Therefore, equation (122-4) gives $-\frac{\pi}{2} + C = 0$ and

$$s \ f(s) = -\tan^{-1} s + \frac{\pi}{2} = \cot^{-1} s \qquad\qquad\text{(122-5)}$$

Hence,

$$f(s) = L\{Si(t)\} = \frac{1}{s}.\cot^{-1} s \qquad\qquad\text{(122-6)}$$

2.18. The Cosine -integral function Ci(t)

This is a tabulated function defined by

$$Ci(t) = -\int_t^\infty \frac{\cos u}{u}.du, \qquad t > 0 \qquad\qquad\text{(123)}$$

Change the variable u into a new variable x, by the relation u = xt, where t is a constant.

Then

$$Ci(t) = -\int_1^\infty \frac{\cos xt}{xt}.t \ dx = -\int_1^\infty \frac{\cos xt}{x} \ dx$$

Therefore,

$$L\{Ci(t)\} = -\int_0^\infty e^{-st} \left\{ \int_1^\infty \frac{\cos xt}{x} \ dx \right\} dt$$

By interchanging the order of integration, we get

$$L\{Ci(t)\} = -\int_1^\infty \frac{1}{x} \left\{ \int_0^\infty e^{-st} \cos xt \ dt \right\} dx$$

$$= -\int_1^\infty \frac{1}{x} \left\{ \frac{s}{s^2 + x^2} \right\} dx = -\int_1^\infty \frac{1}{s} \left\{ \frac{1}{x} - \frac{x}{s^2 + x^2} \right\} dx$$

$$= -\frac{1}{s} \left[\log x - \frac{1}{2}\log(s^2 + x^2) \right]_1^\infty = -\frac{1}{s} \left[\log \frac{x}{\sqrt{s^2 + x^2}} \right]_1^\infty$$

$$= -\frac{1}{s} \left[\log \frac{x}{\sqrt{s^2 + x^2}} \right]_1^\infty = -\frac{1}{s} \left[\log \frac{1}{\sqrt{\left(\frac{s}{x}\right)^2 + 1}} \right]_1^\infty$$

$$= -\frac{1}{s} \left[\log(1) - \log \frac{1}{\sqrt{s^2 + 1}} \right] = -\frac{1}{s} \log \sqrt{s^2 + 1}$$

$$= -\frac{1}{2s} \log(s^2 + 1), \qquad\qquad s > 0 \qquad\qquad\text{(123-1)}$$

The Laplace transform in (123-1) can also be obtained in the following *way:*

$$F(t) = -\int_t^\infty \frac{\cos u}{u}.du \qquad (123\text{-}2)$$

Then

$$\frac{d}{dt}F(t) = \frac{\cos t}{t} \qquad (123\text{-}3)$$

Then, the Laplace transform of (123-3) is

$$-\frac{d}{ds}\left[\; s\; f(s) - F(0)\right] = \frac{s}{s^2+1}$$

Therefore,

$$\frac{d}{ds}\left[\; s\; f(s)\right] = -\frac{s}{s^2+1} \qquad (123\text{-}4)$$

Integrating (123-4), we get

$$s\; f(s) = -\int \frac{sds}{s^2+1} = -\frac{1}{2}\log\!\left(s^2+1\right)+C \qquad (123\text{-}5)$$

And according to the final-value theorem

$$\lim_{s\to\infty} s\, f(s) - F(0) = \lim_{t\to 0} F(t)$$

Therefore,

$$\lim_{s\to 0} s\; f(s) = C$$

Therefore,

$$s\; f(s) = -\frac{1}{2}\log\!\left(s^2+1\right)$$

i.e.,
$$L\{Ci(t)\} = f(s) = -\frac{1}{2s}\log\!\left(s^2+1\right) \qquad (123\text{-}6)$$

2.19. The exponential integral function

The exponential integral function is defined by

$$Ei(t) = \int_{-\infty}^{t} \frac{e^u}{u}.du, \qquad t < 0 \qquad (124)$$

First, substitute by $u = -y$ in equation (124), we get

$$-Ei(-t) = \int_{t}^{\infty} \frac{e^{-y}}{y}.dy, \qquad (124\text{-}1)$$

Second, substitute by $y = tx$, where t is a constant, we get

$$-Ei(-t) = \int_{1}^{\infty} \frac{e^{-tx}}{x}.dx, \qquad t > 0 \qquad (124\text{-}2)$$

The Laplace transform of (124-2) is

$$L\{-Ei(-t)\} = L\left\{\int_{1}^{\infty} \frac{e^{-tx}}{x}.dx\right\} = \int_{0}^{\infty} e^{-st}\left\{\int_{1}^{\infty} \frac{e^{-tx}}{x}.dx\right\}dt \qquad (124\text{-}3)$$

$$= \int_{1}^{\infty} \frac{1}{x}\left\{\int_{0}^{\infty} e^{-st}e^{-tx}.dt\right\}dx = \int_{1}^{\infty} \frac{1}{x}\left(\frac{1}{s+x}\right)dx$$

$$= \frac{1}{s}\int_{1}^{\infty}\left(\frac{1}{x} - \frac{1}{s+x}\right)dx = \frac{1}{s}\left(\log\frac{x}{s+x}\right)_{1}^{\infty}$$

$$= -\frac{1}{s}\left(\log\frac{1}{s+1}\right) = \frac{1}{s}\log(s+1), \qquad s > 0 \qquad (124\text{-}4)$$

2.20. Evaluation of definite integrals using the Laplace transformation

Example 76

Evaluate $\qquad F(t) = \int_{0}^{\infty} \frac{\sin tx}{x(x^2+1)}.dx, \qquad t > a \qquad (125)$

Solution

$$L\{F(t)\} = \int_{0}^{\infty} e^{-st}\left\{\int_{0}^{\infty} \frac{\sin tx}{x(x^2+1)}.dx\right\}dt \qquad (125\text{-}1)$$

Interchanging the order of integration, we get

$$L\{F(t)\} = \int_{0}^{\infty} \frac{1}{x(x^2+1)}\left\{\int_{0}^{\infty} e^{-st}\sin tx.dt\right\}dx$$

$$= \int_0^\infty \frac{1}{x(x^2+1)} \cdot \frac{x}{s^2+x^2} dx = \int_0^\infty \frac{1}{(x^2+1)(s^2+x^2)} dx$$

$$= \int_0^\infty \left(\frac{A}{x^2+1} + \frac{B}{s^2+x^2} \right) dx \tag{125-2}$$

Where the constants A and B are evaluated as usual, as follows

$A(s^2+x^2) + B(x^2+1) = 1$
$A + B = 0$
$As^2 + B = 1$
$As^2 - A = 1$

Therefore,
$A = 1/(s^2-1)$
$B = -1/(s^2-1)$

Thus, equation (125-2) becomes

$$L\{F(t)\} = \frac{1}{s^2-1} \int_0^\infty \left(\frac{1}{x^2+1} - \frac{1}{s^2+x^2} \right) dx$$

$$= \frac{1}{s^2-1} \left(\tan^{-1} s - \frac{1}{s} \tan^{-1} \frac{x}{s} \right)_0^\infty$$

$$= \frac{1}{s^2-1} \left(\frac{\pi}{2} - \frac{1}{s} \cdot \frac{\pi}{2} \right) = \frac{\pi}{2(s^2-1)} \left(1 - \frac{1}{s} \right)$$

$$= \frac{\pi}{2s(s+1)} = \frac{\pi}{2} \left(\frac{1}{s} - \frac{1}{s+1} \right) \tag{125-3}$$

Therefore,

$$F(t) = \frac{\pi}{2} L^{-1} \left\{ \frac{1}{s} - \frac{1}{s+1} \right\} = \frac{\pi}{2} \left(1 - e^{-t} \right) \tag{125-4}$$

Example 77

Evaluate $\quad \int_0^\infty e^{-x^2} dx$ $\qquad\qquad$ (126)

Solution

133

Consider $\qquad F(t) = \int_0^\infty e^{-tx^2} dx,$ $\qquad\qquad\qquad\qquad t > 0$ $\qquad\qquad$ (126-1)

$$L\{F(t)\} = \int_0^\infty e^{-st}\left\{\int_0^\infty e^{-tx^2} dx\right\} dt$$

$$= \int_0^\infty \left\{\int_0^\infty e^{-st}.e^{-tx^2} dt\right\} dx$$

$$= \int_0^\infty \frac{1}{s+x^2} dx = \left[\frac{1}{\sqrt{s}} \tan^{-1}\left(\frac{x}{\sqrt{s}}\right)\right]_0^\infty$$

$$= \int_0^\infty \frac{1}{s+x^2} dx = \frac{1}{\sqrt{s}} \frac{\pi}{2} \qquad\qquad (126-2)$$

The inverse Laplace transform of (126-2) is

$$F(t) = \frac{\pi}{2} L^{-1}\left\{\frac{1}{\sqrt{s}}\right\} = \frac{\pi}{2}\frac{1}{\sqrt{\pi t}} = \frac{1}{2}\sqrt{\frac{\pi}{t}} \qquad\qquad (126-3)$$

Since we introduced the constant t in equation (126-1), we will put t = 1 in (126-3), *we get*

$$F(1) = \int_0^\infty e^{-x^2} dx = \frac{\sqrt{\pi}}{2} \qquad\qquad (126-4)$$

Example 78

Evaluate $\qquad \int_0^t J_0(t - \lambda)J_0(\lambda)d\lambda$ $\qquad\qquad\qquad\qquad$ (127)

Solution

The integral in (127) is the convolution integral encountered in (102-2) denoted as $J_0(t)*J_0(t)$. Also, the Laplace transform of Bessel function was obtained in example (49), equation (84).

From equation (84-3) we get

$$J_0(s) = \frac{1}{\sqrt{(s^2 +1)}}$$

Therefore, the integral of the convolution integral (127) is

$$J_o(s).J_o(s) = \frac{1}{\sqrt{(s^2+1)}} \cdot \frac{1}{\sqrt{(s^2+1)}} = \frac{1}{s^2+1} \qquad (127\text{-}1)$$

Taking the inverse Laplace transform of (127-1), we get

$$\int_0^t J_o(t-\lambda)J_o(\lambda)d\lambda = L^{-1}\left\{\frac{1}{s^2+1}\right\} = \sin t \qquad (127\text{-}2)$$

Example 79

Evaluate $\qquad F(t) = \int_0^\infty \frac{\cos tx}{\sqrt{x}}.dx$, $\qquad\qquad t > o$ $\qquad (128)$

Solution

The Laplace transform of (128) is

$$L\{F(t)\} = \int_0^\infty e^{-st}\left\{\int_0^\infty \frac{\cos tx}{\sqrt{x}}.dx\right\}dt \qquad (128\text{-}1)$$

Interchange the order of integration in (128-1), we get

$$L\{F(t)\} = \int_0^\infty \frac{1}{\sqrt{x}}\left\{\int_0^\infty e^{-st}\cos tx.dt\right\}dx \qquad (128\text{-}2)$$

$$= \int_0^\infty \frac{1}{\sqrt{x}} \cdot \frac{s}{s^2+x^2} dx \qquad (128\text{-}3)$$

We could use the polar coordinates to evaluate the integral (128-3) by putting

$\qquad x = s \tan\theta$
And, as $\qquad x \to \infty$ then $\theta \to \pi/2$
Therefore,
$\qquad dx = s \sec^2\theta. \, d\theta$

Substituting by those polar equivalents in equation (128-3), we get

$$f(s) = \int_0^{\frac{\pi}{2}} \frac{1}{\sqrt{s\tan\theta}} \cdot \frac{s}{s^2\sec^2\theta} \; s \; \sec^2\theta \; d\theta \qquad (128\text{-}4)$$

$$= \frac{1}{\sqrt{s}}\int_0^{\frac{\pi}{2}} \cos^{\frac{1}{2}}\theta\sin^{-\frac{1}{2}}\theta d\theta \qquad (128\text{-}5)$$

135

Equation (128-5) is the **Beta function** β (m, n), which can be evaluated in terms of **Gamma function** Γ(m) and Γ(n) as follows.

****** Start of Auxiliary Proof of Beta Function Property *******

From equation (11), Gamma function is defined as

$$\Gamma\,(n) = \int_0^\infty e^{-x} . x^{n-1} \ dx \tag{128-5.1}$$

We substitute by $x = y^2$ and $dx = 2$ y dy, thus, equation (128-5.1) becomes

$$\Gamma\,(n) = \int_0^\infty e^{-y^2} . y^{2(n-1)} \ 2y.dy = 2\int_0^\infty e^{-y^2} . y^{2n-1} \ dy \tag{128-5.2}$$

Similarly,

$$\Gamma\,(m) = 2\int_0^\infty e^{-x^2} . x^{2n-1} \ dx \tag{128-5.3}$$

From (128-5.2) and (128-5.3), the product Γ(m).Γ(n) can be written as

$$\Gamma\,(m)\Gamma\,(n) = 2.2. \int_{x=0}^\infty \int_{y=0}^\infty e^{-x^2} e^{-y^2} . x^{2n-1} . y^{2m-1} \ dx.dy$$

$$= 4 \int_{x=0}^\infty \int_{y=0}^\infty e^{-(x^2+y^2)} . x^{2n-1} . y^{2m-1} \ dx.dy \tag{128-5.4}$$

We then transform equation (128-5.4) onto the polar coordinate system by the following substitutions:

$$
\begin{aligned}
x \quad &= r \cos \theta \\
y \quad &= r \sin \theta \\
r^2 \quad &= x^2 + y^2 \\
dx \quad &= -r \sin \theta \, d\theta + \cos \theta \, dr \\
dy \quad &= r \cos \theta \, d\theta + \sin \theta \, dr \\
dx.dr \quad &= (-r \sin \theta \, d\theta + \cos \theta \, dr).(\, r \cos \theta \, d\theta + \sin \theta \, dr) \\
&= -r^2 \sin \theta \cos \theta \, (d\theta)^2 - r \sin^2 \theta \, d\theta \, dr \\
&\quad + r \cos^2 \theta \, d\theta \, dr + \cos \theta \sin \theta \, (dr)^2 \tag{128-5.5}
\end{aligned}
$$

We will reasonably ignore the terms of the second orders, namely; $(d\theta)^2$ and $(dr)^2$.

Thus, equation (128-5.5) becomes

$$
\begin{aligned}
dx.dr \quad &= -r \sin^2 \theta \, d\theta \, dr + r \cos^2 \theta \, d\theta \, dr \\
&= (-\sin^2 \theta + \cos^2 \theta) \, d\theta \, dr = (-\sin^2 \theta + 1 - \sin^2 \theta) r \, d\theta \, dr
\end{aligned}
$$

$$= r \, d\theta \, dr \qquad (128\text{-}5.6)$$

Therefore, the polar form of equation (128-5.4) becomes

$$\Gamma(m).\Gamma(n) == 4 \int_{r=0}^{\infty} \int_{\theta=0}^{\frac{\pi}{2}} e^{-r^2}.r^{2n+2m-2}\cos^{2n-1}\theta.\sin^{2m-1}\theta \; r.dr.d\theta \qquad (128\text{-}6)$$

We will arrange the independent integrals in (128-6) in order to separate the Gamma functions from the remainder of the terms in that equation as follows.

$$\Gamma(m).\Gamma(n) == 2 \int_{r=0}^{\infty} e^{-r^2}.r^{2n+2m-1}.dr \int_{\theta=0}^{\frac{\pi}{2}} 2.\cos^{2n-1}\theta.\sin^{2m-1}\theta \,.d\theta \qquad (128\text{-}6.1)$$

Comparing the first integral in (128-6.1) with the Gamma integral in (128-5.2), we can now write equation (128-6.1) in terms of $\Gamma(m)$, $\Gamma(n)$, and $\Gamma(m+n)$, as follows

$$\Gamma(m).\Gamma(n) = \Gamma(m+n) \int_{\theta=0}^{\frac{\pi}{2}} 2.\cos^{2n-1}\theta.\sin^{2m-1}\theta \,.d\theta \qquad (128\text{-}6.2)$$

Hence, the remaining integral in (128-6.2) is given as the Beta integral of the form sought after in equation (128-5), as follows

$$\beta(m,n) = \frac{\Gamma(m).\Gamma(n)}{\Gamma(m+n)} == 2 \int_{\theta=0}^{\frac{\pi}{2}} \cos^{2n-1}\theta.\sin^{2m-1}\theta \,.d\theta \qquad (128\text{-}7)$$

From equation (128-7), we can write (128-5) as follows

$$f(s) = \int_{0}^{\infty} \frac{1}{\sqrt{x}}.\frac{s}{s^2+x^2}dx = \frac{1}{2\sqrt{s}}\left(2.\int_{0}^{\frac{\pi}{2}} \cos^{\frac{1}{2}}\theta \sin^{-\frac{1}{2}}\theta d\theta\right)$$

$$= \frac{1}{2\sqrt{s}}.\beta\left(\frac{1}{4},\frac{3}{4}\right) = \frac{1}{2\sqrt{s}}.\frac{\Gamma\left(\frac{1}{4}\right).\Gamma\left(\frac{3}{4}\right)}{\Gamma(1)} \qquad (128\text{-}8)$$

The determination of isolated $\Gamma\left(\frac{1}{4}\right)$ or $\Gamma\left(\frac{3}{4}\right)$ requires numerical computation, but the product $\Gamma\left(\frac{1}{4}\right).\Gamma\left(\frac{3}{4}\right)$ can be determined from equation (128-6.2) as follows.

$$\Gamma\left(\frac{1}{4}\right).\Gamma\left(\frac{3}{4}\right) = \Gamma(\frac{1}{4}+\frac{3}{4}) \int_{\theta=0}^{\frac{\pi}{2}} 2.\cos^{2\left(\frac{1}{4}\right)-1}\theta.\sin^{2\left(\frac{3}{4}\right)-1}\theta \,.d\theta \qquad (128\text{-}8.1)$$

$$= \Gamma(1) \int_{\theta=0}^{\frac{\pi}{2}} 2.\cos^{-\frac{1}{2}}\theta.\sin^{\frac{1}{2}}\theta\,.d\theta$$

$$= 2 \int_{\theta=0}^{\frac{\pi}{2}} \cot^{\frac{1}{2}}\theta.d\theta \qquad\qquad\qquad (128\text{-}8.2)$$

Integrating by parts, (128-8.2) gives

$$2 \int_{\theta=0}^{\frac{\pi}{2}} \cot^{\frac{1}{2}}\theta.d\theta = \frac{1}{\sqrt{2}}\left\{ \log\frac{\cot\theta + \sqrt{2\cot\theta} + 1}{\cot\theta - \sqrt{2\cot\theta} + 1} + 2\left[\tan^{-1}\left(1 - \sqrt{2\cot\theta}\right) - \tan^{-1}\left(1 + \sqrt{2\cot\theta}\right)\right] \right\}_{0}^{\frac{\pi}{2}}$$

$$= \frac{1}{\sqrt{2}}\left\{ \left(\log\frac{0+1}{0+1} + 2\left[\tan^{-1}(1-0) - \tan^{-1}(1+0)\right]\right) - \log\frac{1+0}{1-0} - 2\left[\tan^{-1}(-\infty) - \tan^{-1}(\infty)\right] \right\}$$

$$= \frac{1}{\sqrt{2}}\left\{ - 2\left[\tan^{-1}(-\infty) - \tan^{-1}(\infty)\right] \right\}$$

$$= \frac{1}{\sqrt{2}}\left\{ -2\left[\frac{-\pi}{2} - \frac{\pi}{2}\right] \right\} = \frac{\pi}{\sqrt{2}} \qquad\qquad (128\text{-}8.3)$$

In fact, with some laborious derivation, we could prove that

$$\Gamma(m).\Gamma(1-m) = \frac{\pi}{\text{sim}(m\pi)} \qquad\qquad \text{for } 0 < m < 1 \qquad (128\text{-}8.4)$$

****** End of Auxiliary Proof of Beta Function Property *******

Therefore, our initial integral in equation (128-5) becomes

$$f(s) = \int_{0}^{\infty} \frac{1}{\sqrt{x}}.\frac{s}{s^2 + x^2}\,dx = \frac{1}{\sqrt{s}}\int_{0}^{\frac{\pi}{2}} \cos^{\frac{1}{2}}\theta \sin^{-\frac{1}{2}}\theta\, d\theta$$

$$= \frac{1}{2\sqrt{s}}.\frac{\pi}{\sin\frac{\pi}{4}} = \frac{1}{2\sqrt{s}}.\frac{\pi}{\sqrt{\frac{1}{2}}} = \frac{1}{\sqrt{2}}.\frac{\pi}{\sqrt{s}} \qquad\qquad (128\text{-}9)$$

The inverse transform of (128-9) is thus

$$F(t) = \int_{0}^{\infty} \frac{\cos tx}{\sqrt{x}}.dx = L^{-1}\left\{ \frac{1}{\sqrt{2}}.\frac{\pi}{\sqrt{s}} \right\} = \frac{1}{\sqrt{2}}.\frac{\pi}{\sqrt{\pi t}} = \sqrt{\frac{\pi}{2t}} \qquad\qquad (128\text{-}10)$$

Example 80

Evaluate $\qquad F(t) = \int_0^\infty \sin x^2 .dx$ $\qquad\qquad\qquad\qquad$ (129)

Solution

Let $\qquad F(t) = \int_0^\infty \sin tx^2 .dx ,$ $\qquad\qquad t > 0$ $\qquad\qquad$ (129-1)

The Laplace transform of (129-1) is

$$L\{F(t)\} = \int_0^\infty e^{-st} .\left\{\int_0^\infty \sin tx^2 .dx\right\} dt \qquad\qquad (129\text{-}2)$$

Interchanging the order of integration in (129-2) gives

$$L\{F(t)\} = \int_0^\infty \left\{\int_0^\infty e^{-st} \sin tx^2 .dt\right\} dx$$

$$= \int_0^\infty \frac{x^2}{s^2 + x^4} dx \qquad\qquad\qquad (129\text{-}3)$$

We will convert the coordinates in (129-3) into the polar form by putting

$$x^2 - s \tan\theta$$
$$2x\, dx = s \sec^2\theta\, d\theta$$

When

$$x \to 0 , \theta \to 0$$
$$x \to \infty , \theta \to \pi/2$$

Therefore, equation (129-3) becomes

$$\int_0^\infty \frac{x^2}{s^2 + x^4} dx = \int_0^\infty \frac{s\tan\theta}{s^2 + (s\tan\theta)^2} . \frac{s\,\sec^2\theta}{2\sqrt{s\tan\theta}} d\theta$$

$$= \frac{1}{2}\int_0^\infty \frac{\sqrt{s\tan\theta}\,\,s\,\sec^2\theta}{s^2 \sec^2\theta} d\theta = \frac{1}{2\sqrt{s}}\int_0^\infty \sqrt{\tan\theta} .d\theta \qquad (129\text{-}4)$$

139

We have already dealt with similar integral in equation (128-7) which resulted into equation (128-8). Thus, equation (129-4) can be evaluated as before, as follows.

$$\int_0^\infty \frac{x^2}{s^2+x^4} dx = \frac{1}{2\sqrt{s}} \int_0^\infty \sqrt{\tan\theta}.d\theta = \frac{1}{2\sqrt{s}} \int_0^\infty \sin^{\frac{1}{2}}\theta.\cos^{-\frac{1}{2}}\theta.d\theta \qquad (129\text{-}5)$$

$$= \frac{1}{4\sqrt{s}} \int_0^\infty 2\sin^{\frac{1}{2}}\theta.\cos^{-\frac{1}{2}}\theta.d\theta = \frac{1}{4\sqrt{s}} \beta\left(\frac{1}{4},\frac{3}{4}\right)$$

$$= \frac{1}{4\sqrt{s}} \left[\frac{\Gamma\left(\frac{1}{4}\right).\Gamma\left(\frac{3}{4}\right)}{\Gamma(1)} \right]$$

$$= \frac{1}{4\sqrt{s}} \left(\frac{\pi}{\sin\frac{\pi}{4}} \right) = \frac{\pi}{2\sqrt{2s}} \qquad (129\text{-}5)$$

Laplace transform inverting (129-5), we get

$$F(t) = \int_0^\infty \sin tx^2.dx = L^{-1}\left\{ \frac{\pi}{2\sqrt{2s}} \right\} = \frac{\pi}{2\sqrt{2}}.\frac{1}{\sqrt{\pi t}} = \frac{\sqrt{\pi}}{2\sqrt{2t}} \qquad (129\text{-}6)$$

Put t = 1 in (129-6), we get the required evaluation as follows

$$F(t) = \int_0^\infty \sin x^2.dx = \frac{1}{2}\sqrt{\frac{\pi}{2}} \qquad (129\text{-}7)$$

2.21. The Heaviside's expansion formulae

Let
$$f(s) = \frac{p(s)}{q(s)} \qquad (130)$$

where p(s) and q(s) are polynomials in s, p(s) being of lower degree than q(s).

Case I   **All factors of q(s) are linear and distinct**

Consider the case in which all factors of q(s) are linear and distinct.

Let
$$q(s) = (s-\alpha_1)(s-\alpha_2)(s-\alpha_3)...(s-\alpha_n) \qquad (131)$$

α 's being all distinct

Then
$$f(s) = \frac{p(s)}{q(s)} = \frac{A_1}{s - \alpha_1} + \frac{A_2}{s - \alpha_2} + .. + \frac{A_r}{s - \alpha_r} + .. = \frac{A_n}{s - \alpha_n} \qquad (132\text{-}1)$$

Clearing fractions, we get

$$\begin{aligned}
p(s) \quad &= A_1 \, (s - \alpha_2)\, (s - \alpha_3)... \, (s - \alpha_n) \\
&+ A_2 \, (s - \alpha_1)\, (s - \alpha_3)... \, (s - \alpha_n) \\
&+ .... \\
&+ A_r \, (s - \alpha_1)\, (s - \alpha_3)... \, (s - \alpha_{r\text{-}1})\, (s - \alpha_{r+1})....\, (s - \alpha_n) \\
&+ .... \\
&+ A_n \, (s - \alpha_1)\, (s - \alpha_3)... \, (s - \alpha_{n\text{-}1})
\end{aligned} \qquad (132\text{-}2)$$

The various coefficients A_r can be evaluated by substituting by $s = \alpha_r$ in (132-2) as follows

$$p(\alpha_1) = A_1 \, (s - \alpha_2)\, (s - \alpha_3)... \, (s - \alpha_n) \qquad (132\text{-}3)$$

Since all terms in (132-2) that contain $(s - \alpha_1)$ vanish. Therefore,

$$A_1 = \frac{p(\alpha_1)}{(\alpha_1 - \alpha_2)(\alpha_1 - \alpha 3)...(\alpha_1 - \alpha_n)} \qquad (132\text{-}4)$$

Hence, A_1 is obtained by substituting α_1 in the numerator and in all factors of the denominator except $s - \alpha_1$ itself. Similarly for A_2, A_3, ..., A_n.

Now, differentiating (131) with respect to s, we get

$$\begin{aligned}
q'(s) \quad &= (s - \alpha_2)\, (s - \alpha_3)... \, (s - \alpha_n) \\
&+ (s - \alpha_1)\, (s - \alpha_3)... \, (s - \alpha_n) \\
&+.... \\
&+ (s - \alpha_1)\, (s - \alpha_3)... \, (s - \alpha_{n\text{-}1})
\end{aligned} \qquad (132\text{-}5)$$

Therefore, upon substituting by $s = \alpha_1$, in 9132-5), we get

$$q'(\alpha_1) \quad = (s - \alpha_2)\, (s - \alpha_3)... \, (s - \alpha_n) \qquad (132\text{-}6)$$

From (132-6) we can write (132-4) as follows

$$A_1 - \frac{p(\alpha_1)}{q'(\alpha_1)}$$

In general,

$$A_r = \frac{p(\alpha_r)}{q'(\alpha_r)} \qquad (132\text{-}7)$$

Therefore, from (132-7), the fraction in (130) can be written as

$$f(s) = \frac{p(s)}{q(s)} = \sum_{r=1}^{n} \frac{p(\alpha_r)}{q'(\alpha_r)} \cdot \frac{1}{s - \alpha_r} \tag{133}$$

Therefore, the inverse Laplace transform of (130) is given by

$$F(t) = L^{-1}\left\{\frac{p(s)}{q(s)}\right\} = \sum_{r=1}^{n} \frac{p(\alpha_r)}{q'(\alpha_r)} \cdot e^{\alpha_r t} \tag{134}$$

Case II   **The denominator, g(s) has repeated linear factors**

Suppose that the denominator $q(s)$ has a repeated linear factor $(s - \alpha)^n$ and that

$$f(s) = \frac{\phi(s)}{(s - \alpha)^n} \tag{135}$$

Where $\varphi(s)$ consists of the numerator p(s) and **all** factors of q(s) **except** $(s - \alpha)^n$.

Therefore,

$$\phi(s) = \phi(\alpha + \overline{s - \alpha}) = \phi(\alpha) + (s - \alpha)\phi'(\alpha) + \frac{(s - \alpha)^2}{2!}\phi''(\alpha) + ..$$

$$+ \frac{(s - \alpha)^{n-1}}{(n-1)!}\phi^{(n-1)}(\alpha) + (s - \alpha)^n h(\alpha) \tag{136}$$

Therefore,

$$\frac{p(s)}{q(s)} = \frac{\phi(s)}{(s - \alpha)^n} = \frac{\phi(\alpha)}{(s - \alpha)^n} + \frac{\phi'(\alpha)}{(s - \alpha)^{n-1}} + \frac{\phi''(\alpha)}{2!(s - \alpha)^{n-2}} + .. + \frac{\phi^{(n-1)}(\alpha)}{(n-1)!(s - \alpha)} + h(s) \tag{137}$$

Therefore, the inverse Laplace transform of the fraction (137) is

$$L^{-1}\left\{\frac{p(s)}{q(s)}\right\} = e^{\alpha t}\left[\frac{t^{n-1}}{(n-1)!}\frac{\phi(\alpha)}{0!} + \frac{t^{n-2}}{(n-2)!}\frac{\phi'(\alpha)}{1!} + \frac{t^{n-3}}{(n-3)!}\frac{\phi''(\alpha)}{2!} + .. + \frac{\phi^{(n-1)}(\alpha)}{(n-1)!}\right] + H(t)$$

$$= e^{\alpha t}\sum_{r=1}^{n} \frac{t^{n-r}}{(n-r)!}\frac{\phi^{(r-1)}(\alpha)}{(r-1)!} \tag{138}$$

Case III: **q(s) has quadratic factors**

142

Let $\quad f(s) = \dfrac{p(s)}{q(s)} = \dfrac{\phi(s)}{(s-\alpha)^2 + b^2}$ $\hfill$ (139)

Where $\phi(s)$ consists of p(s) and **all** factors of q(s) **except** $(s-\alpha)^2 + b^2$ itself.

Let $\quad \dfrac{\phi(s)}{(s-\alpha)^2 + b^2} = \dfrac{As + B}{(s-\alpha)^2 + b^2} + h(s)$ $\hfill$ (139-1)

Therefore, $\quad \phi(s) = As + B + [(s-\alpha)^2 + b^2]\, h(s)$ $\hfill$ (139-2)

Put $\quad s = \alpha + i\, b$ $\hfill$ (139-3)

Thus, $\quad \phi(\alpha + i\, b) = A(\alpha + i\, b) + B$ $\hfill$ (139-4)

Let $\quad \phi(\alpha + i\, b) = \phi_1 + i\, \phi_2$ $\hfill$ (139-5)

Therefore, $\quad \phi_1 + i\, \phi_2 = A\alpha + B + i\, b\, A$

$\quad \phi_1 = A\alpha + B$
$\quad \phi_2 = b\, A$

Therefore,

$\quad A = \phi_2 / b$
$\quad B = (b\, \phi_1 - \alpha\, \phi_2) / b$ $\hfill$ (139-6)

Substituting by A and B from (139-6) into (139-1), we get

$$\dfrac{\phi(s)}{(s-\alpha)^2 + b^2} = \dfrac{\dfrac{\phi_2}{b}s + \dfrac{b\phi_1 - \alpha\phi_2}{b}}{(s-\alpha)^2 + b^2} + h(s)$$

$$= \dfrac{1}{b}\left(\dfrac{\phi_2 s + b\phi_1 - \alpha\phi_2}{(s-\alpha)^2 + b^2}\right) + h(s)$$

$$- \dfrac{1}{b}\left(\dfrac{\phi_2(s-\alpha) + b\phi_1}{(s-\alpha)^2 + b^2}\right) + h(s) \hfill (139\text{-}7)$$

Therefore, the inverse Laplace transform of (139) becomes

$$L^{-1}\left\{\dfrac{p(s)}{q(s)}\right\} = L^{-1}\left\{\dfrac{\phi(s)}{(s-\alpha)^2 + b^2}\right\} = \dfrac{1}{b}L^{-1}\left(\dfrac{\phi_2(s-\alpha) + b\phi_1}{(s-\alpha)^2 + b^2}\right) + L^{-1}\{h(s)\}$$

$$= \frac{1}{b}e^{at}\left(\phi_2 \cos bt + \phi_1 \sin bt\right) + H(t) \tag{140}$$

Case IV   **q(s) has repeated quadratic factors**

Consider for example the particular case when

Let $f(s) = \dfrac{p(s)}{q(s)} = \dfrac{\phi(s)}{\left[(s-\alpha)^2 + b^2\right]^2} = \dfrac{As+B}{\left[(s-\alpha)^2 + b^2\right]^2} + \dfrac{Cs+D}{(s-\alpha)^2 + b^2} + h(s)$ (141)

Therefore, $\phi(s) = As + B + (Cs + D)[(s-\alpha)^2 + b^2] + h(s)\,[(s-\alpha)^2 + b^2]^2$ (141-1)

The four unknowns A, B, C, and D can be obtained in the same manner we did in (139-3) thru (139-6) by substituting $s = \alpha + i\,b$ in (141-1) and in the equation derived by differentiating (141-1) w.r.t. s and equating real to real and imaginary to imaginary on both sides in each case.

2.22. The inversion integral

The inversion integral is a powerful as well as a direct mean for finding inverse Laplace transforms.

We have so far assumed s to be real. We shall now consider s to be complex.
Let F(t) be a real function of the positive, real variable t, sectionally continuous in each finite interval $0 \le t \le T$ and of exponential order as $t \to \infty$

$$f(s) = \int_0^\infty e^{-st}F(t)dt, \qquad \text{where } s = x + i\,y \tag{142}$$

Then the inversion integral formula is given by

$$F(t) = \frac{1}{2\pi i} \lim_{\beta \to \infty} \int_{\gamma - i\beta}^{\gamma + i\beta} e^{st}f(s)ds, \tag{143}$$

This formula is also known as **Bromwich's integral formula**, and it is a modified form of the **Fourier integral formula**. The integration is carried in the complex plane of s along the line x = y which is taken far enough to the right such that all poles, branch points or essential singularities of e^{st} f(s) lie to the left of it. In practice, the integral (143) is evaluated by considering first the contour integral

$$\frac{1}{2\pi i} \int_C e^{st}f(s)ds$$

Where, C is the contour consisting of:

(i) The line AB whose equation is x = y and

144

(ii) The circular arc BDA which we denote by Γ of a circle centre O and radius R. R is taken large enough such that all poles of e^{st} f(s) lie inside C, and in the limit R is made infinite.

Now, since $\beta = \sqrt{R^2 - \gamma^2}$ hence when $\beta \to \infty$ then $R \to \infty$ and according to the inversion formula, equation (143) becomes

$$F(t) = \frac{1}{2\pi i} \lim_{R \to \infty} \int_{\gamma - i\beta}^{\gamma + i\beta} e^{st} f(s) ds = \frac{1}{2\pi i} \lim_{R \to \infty} \left\{ \int_C e^{st} f(s) ds - \int_{\Gamma} e^{st} f(s) ds \right\} \qquad (143\text{-}1)$$

Assuming that all singularities of f(s) are poles and that

$$\int_{\Gamma} e^{st} f(s) ds \to 0 \quad \text{as } R \to \infty \qquad (143\text{-}2)$$

Then, we shall show that the Cauchy's theorem of residues requires that

$$\int_C e^{st} f(s) ds = 2\pi i \qquad (143\text{-}3)$$

(sum of residues of e^{st} f(s) at its poles inside C)

It follows that

$$F(t) = \textbf{sum of residues of } e^{st} \text{ f(s) at poles of f(s).} \qquad (143\text{-}4)$$

********** Proof of Cauchy's Theorem of Residues ************

2.22.1. Cauchy's Theorem of Residue

The **line-integral** of a function, that is regular in a domain D, which is enclosed by a closed contour C and contains a finite numbers of poles $a, a', a'', ..,$ in its interior, is given by:

$$\oint_C f(z) \, dz = 2\pi i \cdot ((\text{sum of residues of } f(z) \text{ at its poles inside C}))$$

Consider the poles of $f(z)$ at $z = 0$ and suppose that $f(z)$ can be expanded around a in the form:

$$f(z) = \frac{C_n}{(z-a)^n} + \frac{C_{n-1}}{(z-a)^{n-1}} + \ldots \ldots + \frac{C_1}{z-a} + \Phi(z) \qquad (143\text{-}4.1)$$

Where $\Phi(z)$ is regular in the same circular domain, with center a and radius r, and similarly for the poles $a, a', a'', ..,$ etc.

145

Now, by taking r, r', r'', .., small enough, we can ensure that they do not overlap and that they all lie inside C.

Draw circle γ, of radius ρ, around a as a center (a is complex variable, not a scalar, see figure) such that $\rho < r$. Further, draw similar circles around the remaining poles.

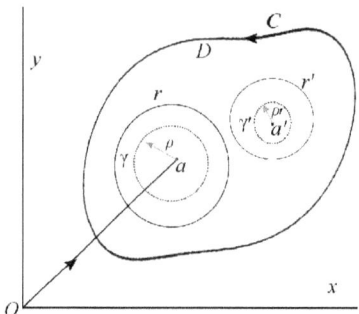

According to Cauchy's Theorem Corollary

$$\oint_C f(z)dz = \oint_\gamma f(z)dz + \oint_{\gamma'} f(z)dz + \,..... \tag{143-4.2}$$

Where all integrals are with positive signs, i.e., counter clockwise.

Now, consider $\oint_\gamma f(z)dz$

$$\oint_\gamma f(z)\,dz = \sum_{m=1}^{n} C_m \int_\gamma \frac{dz}{(z-a)^m} + \int_\gamma \Phi(z)dz \tag{143-4.3}$$

Now, the last integral, $\int_\gamma \Phi(z)dz = 0$ since $\Phi(z)$ is regular for all points on the circle γ.

As illustrated in the figure, below, we can put

$$z = \rho\, e^{i\theta} + a$$
$$dz = i\,\rho\, e^{i\theta}\, d\theta$$

146

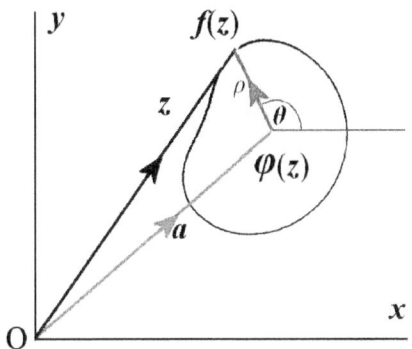

$$\oint_\gamma \frac{dz}{(z-a)^m} = \int_{-\pi}^{\pi} \frac{i\rho e^{i\theta} d\theta}{\rho^m e^{im\theta}}$$

$$= \frac{i}{\rho^{m-1}} \frac{\left[e^{i(1-m)\theta}\right]_{-\pi}^{\pi}}{i(1-m)}$$

$$= \frac{1}{\rho^{m-1}(1-m)}\left[e^{i(1-m)\pi} - e^{-i(1-m)\pi}\right] = 0 \qquad (143\text{-}4.4)$$

Since, $e^{i\theta} - e^{-i\theta} = 2\,i\sin\theta$

When m = 1 (one pole of first order), then

$$\oint_\gamma f(z)\,dz = C_1 \oint_\gamma \frac{dz}{z-a} = C_1\,2\pi i \qquad (143\text{-}4.5)$$

Because,

$$\oint_\gamma f(z)\,dz = \int_{-\pi}^{\pi} \frac{i\rho e^{i\theta} d\theta}{\rho e^{i\theta}} = i\int_{-\pi}^{\pi} d\theta = 2\pi i$$

Similarly, for integrals around γ', γ'', ..

$$\oint_C f(z)\,dz \qquad = 2\pi\,i\,(C_1 + C_2 + C_3 + \ldots)$$

$$= 2\pi\,i\,(\text{sum of residues}) \qquad (143\text{-}4.6)$$

********** End of Proof of Cauchy's Theorem of Residues *************

The result in (143-4) was obtained on the assumption made in equation (143-2). This assumption is satisfied in almost all physical problems. In fact, a sufficient condition for the ultimate vanishing of the above integral is the following.

147

If we can find two positive constants M and k such that on Γ, where $s = Re^{i\theta}$

$$\left| f(s) \right| < \frac{M}{R^k}$$

Then, the validity of (143-2)

2.23. Formulae for residues

I. Simple poles

Let $\qquad \dfrac{p(s)}{q(s)} = \dfrac{\phi(s)}{s - s_o}$ $\hfill$ (144)

Where, $\phi(s)$ consists of $p(s)$ and **all** factors of $q(s)$ **except** the factor $(s - s_o)$ itself. Here, we have a simple polo at $(s - s_o)$.

Now, from equation (136)

$$\phi(s) = \phi(s_o + \overline{s - s_o}) = \phi(s_o) + (s - s_o)\phi'(s_o) + \frac{(s - s_o)^2}{2!}\phi''(s_o) + .. \qquad (144\text{-}1)$$

Therefore,

$$\frac{\phi(s)}{s - s_o} = \frac{\phi(s_o)}{s - s_o} + \phi'(s_o) + \frac{s - s_o}{2!}\phi''(s_o) + .. \qquad (144\text{-}2)$$

Therefore, Residue at $(s = s_o)$ is $\phi(s_o)$

$$\phi(s_o) = \lim_{s \to s_o}(s - s_o)\frac{\phi(s)}{s - s_o} = \lim_{s \to s_o}(s - s_o)\frac{p(s)}{q(s)} \qquad (144\text{-}3)$$

II. Multiple poles

Let $\qquad \dfrac{p(s)}{q(s)} = \dfrac{\phi(s)}{(s - s_o)^m}$ $\hfill$ (145)

Hence, (144-1) becomes

$$\phi(s) = \phi(s_o) + (s - s_o)\phi'(s_o) + \frac{(s - s_o)^2}{2!}\phi''(s_o) + .. + \frac{(s - s_o)^{m-1}}{(m-1)!}\phi^{(m-1)}(s_o) + ... \qquad (145\text{-}1)$$

Equation (145) becomes

$$\frac{p(s)}{q(s)} = \frac{\phi(s)}{(s-s_o)^m} = \frac{\phi(s_o)}{(s-s_o)^m} + \frac{\phi'(s_o)}{(s-s_o)^{m-1}} + \frac{(s-s_o)^2}{2!}\phi''(s_o) + .. + \frac{1}{s-s_o}\frac{\phi^{(m-1)}(s_o)}{(m-1)!} + ... \quad (145\text{-}2)$$

Therefore, Residue at s = s_o of multiplicity m is given by

$$\frac{1}{m-1}\phi^{(m-1)}(s_o) = \frac{1}{(m-1)!}\left[\frac{d^{m-1}}{ds^{m-1}}\phi(s)\right]_{s=s_o} \quad (145\text{-}3)$$

Example 80

Find $L^{-1}\left\{\dfrac{1}{(s-1)(s-2)(s-3)}\right\}$ \quad (146)

Solution

This is equal to the sum of residues of

$$\frac{e^{st}}{(s-1)(s-2)(s-3)} \quad (146\text{-}1)$$

At its poles.

<u>s = 1</u>

$$\text{Re sidue} = \lim_{s\to 1}(s-1)\frac{e^{st}}{(s-1)(s-2)(s-3)} = \lim_{s\to 1}\frac{e^{st}}{(s-2)(s-3)} = \frac{e^t}{2} \quad (146\text{-}2)$$

<u>s = 2</u>

$$\text{Re sidue} = \lim_{s\to 2}(s-2)\frac{e^{st}}{(s-1)(s-2)(s-3)} = \lim_{s\to 2}\frac{e^{st}}{(s-1)(s-3)} = -e^{2t} \quad (146\text{-}3)$$

<u>s = 3</u>

$$\text{Re sidue} = \lim_{s\to 3}(s-3)\frac{e^{st}}{(s-1)(s-2)(s-3)} = \lim_{s\to 3}\frac{e^{st}}{(s-1)(s-2)} = -\frac{e^{wt}}{2} \quad (146\text{-}4)$$

Therefore, from (146-2), (146-3) and (146-4), we get

$$L^{-1}\left\{\frac{1}{(s-1)(s-2)(s-3)}\right\} = \frac{1}{2}e^t - e^{2t} + \frac{1}{2}e^{3t} \quad (146\text{-}5)$$

Example 81

Find $\qquad L^{-1}\left\{\dfrac{1}{s^2(s+1)}\right\}$ $\qquad\qquad$ (147)

Solution

This is equal to the sum of residues of

$$\dfrac{e^{st}}{s^2(s+1)} \qquad\qquad (147\text{-}1)$$

At its poles.

s = 0

The residue at the double pole at s = 0 is

$$\left\{\dfrac{d}{ds}\dfrac{e^{st}}{(s+1)}\right\}_{s=0} = \left\{\dfrac{(s+1)te^{st}-e^{st}}{(s+1)^2}\right\}_{s=0} = t-1 \qquad\qquad (147\text{-}2)$$

s = -1

The residue at s = -1 is

$$\lim_{s\to -1}\dfrac{e^{st}}{s^2} = e^{-t} \qquad\qquad (147\text{-}3)$$

Therefore, from (147-2) and (147-3), we get

$$L^{-1}\left\{\dfrac{1}{s^2(s+1)}\right\} = e^{-t}+t-1 \qquad\qquad (147\text{-}4)$$

Example 82

Find $\qquad L^{-1}\left\{\dfrac{1}{s^2(s^2+\omega^2)}\right\}$ $\qquad\qquad$ (148)

Solution

This is equal to the sum of residues of

$$\dfrac{e^{st}}{s^2(s-i\omega)(s+i\omega)} \qquad\qquad (148\text{-}1)$$

At its poles.

$s = 0$

Residue at the double pole at $s = 0$ is

$$\left[\frac{d}{ds}\frac{e^{st}}{(s^2+\omega^2)}\right]_{s=0} = \left[\frac{(s^2+\omega^2)te^{st}-2se^{st}}{(s^2+\omega^2)^2}\right]_{s=0} = \frac{t}{\omega^2} \qquad (148\text{-}2)$$

$s = \pm i\omega$

Sum of residues at $s = \pm i\omega$ is

$$\left[\frac{e^{st}}{s^2(s+i\omega)}\right]_{s=i\omega} + \left[\frac{e^{st}}{s^2(s-i\omega)}\right]_{s=-i\omega} = -\frac{1}{\omega^3}\left[\frac{e^{i\omega t}-e^{-i\omega t}}{2i}\right] = -\frac{\sin\omega t}{\omega^3} \qquad (148\text{-}3)$$

Therefore, from (148-2) and (148-3), we get

$$L^{-1}\left\{\frac{1}{s^2(s^2+\omega^2)}\right\} = \frac{1}{\omega^3}(\omega t - \sin\omega t) \qquad (148\text{-}4)$$

2.24. Inversion in the case of branch points

Example 83

Find $\qquad L^{-1}\left\{\frac{1}{\sqrt{s}}\right\} \qquad (149)$

Solution

The path of integration of the inversion integral has to be modified if the integrand has a branch point. This is illustrated by the following example.

Suppose it is required to find the inverse transform of $\dfrac{1}{\sqrt{s}}$. Here the function $\dfrac{1}{\sqrt{s}}$ has a branch point at the origin. We therefore cut the s-plane along the negative x-axis and take the path shown in this diagram.

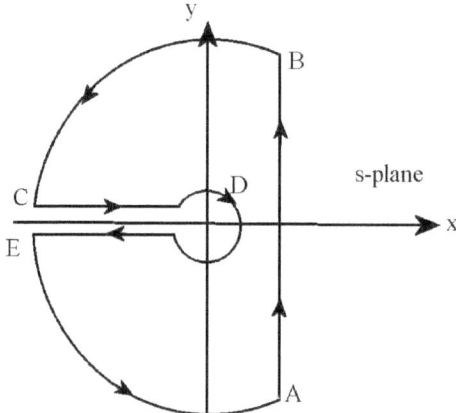

Since $\dfrac{e^{st}}{\sqrt{s}}$ has no singularities within this closed path, hence according to **Cauchy's theorem** we have:

$$\int_{AB} \frac{e^{st}\,ds}{\sqrt{s}} + \int_{BC+EA} \frac{e^{st}\,ds}{\sqrt{s}} + \int_{CDE} \frac{e^{st}\,ds}{\sqrt{s}} = 0 \tag{149-1}$$

It is easy to show that the integral along the circular path of infinite radius is zero and hence, in the limit and when the radius of the circular path is infinite, we get

$$\int_{AB} \frac{e^{st}\,ds}{\sqrt{s}} = -\int_{CDE} \frac{e^{st}\,ds}{\sqrt{s}} = \int_{EDC} \frac{e^{st}\,ds}{\sqrt{s}} \tag{149-2}$$

Now integral on EDC consists of three puts, one around the small circle at D and the others along the straight paths ED and DC.

We shall now evaluate each of the three integrals.

(i) Consider first the integral around the small circle at D. Taking the radius of this small circle to be equal r, then we can put for points on the circle:

$$s = r\,e^{i\theta}$$
$$ds = ir\,e^{i\theta}\,d\theta$$
$$s^{\frac{1}{2}} = r^{\frac{1}{2}}e^{\frac{1}{2}i\theta}$$

Therefore,

$$\int_{\text{small circle D}} \frac{e^{st} ds}{\sqrt{s}} = i \int_{r \to 0} e^{rt(\cos\theta + i\sin\theta)} r^{\frac{1}{2}} e^{i\frac{\theta}{2}} d\theta = 0 \tag{149-3}$$

(ii) Next, consider the integral along ED and put for points on ED

$$s = x\, e^{-i\pi} = -x$$
$$ds = -dx$$
$$s^{\frac{1}{2}} = x^{\frac{1}{2}} e^{-i\frac{\pi}{2}} = -i\sqrt{x}$$

Therefore,

$$\int_{ED} \frac{e^{st} ds}{\sqrt{s}} = -i \int_{\infty}^{0} x^{-\frac{1}{2}} e^{-xt} dx = i \int_{0}^{\infty} x^{-\frac{1}{2}} e^{-xt} dx \tag{149-4}$$

(iii) Finally along DC, put

$$s = x\, e^{i\pi} = -x$$
$$ds = -dx$$
$$s^{\frac{1}{2}} = x^{\frac{1}{2}} e^{i\frac{\pi}{2}} = i\sqrt{x}$$

Therefore,

$$\int_{DC} \frac{e^{st} ds}{\sqrt{s}} = -i \int_{0}^{\infty} x^{-\frac{1}{2}} e^{-xt} dx \tag{149-5}$$

Hence the sum of the three integrals is

$$2i \int_{0}^{\infty} x^{-\frac{1}{2}} e^{-xt} dx \tag{149-6}$$

Put

$$xt = \lambda^2$$
$$t\, dx = 2\lambda\, d\lambda$$

Where, t is constant

Therefore,

$$2i \int_{0}^{\infty} x^{-\frac{1}{2}} e^{-xt} dx = 2i \int_{0}^{\infty} \frac{t^{\frac{1}{2}}}{\lambda} e^{-\lambda^2} \left(\frac{2\lambda}{t}\right) d\lambda$$

$$= \frac{4i}{\sqrt{t}} \int_{0}^{\infty} e^{-\lambda^2} d\lambda = \frac{4i}{\sqrt{t}} \cdot \frac{\sqrt{\pi}}{2} = 2i\sqrt{\frac{\pi}{t}} \tag{149-7}$$

Hence, the Inverse Laplace transform in (149) becomes

153

$$L^{-1}\left\{\frac{1}{\sqrt{s}}\right\} = \frac{1}{2\pi i}\left(2i\sqrt{\frac{\pi}{t}}\right) = \sqrt{\frac{1}{t\pi}} \qquad (150)$$

2.25. Review Examples on Laplace Transform

Example 84

If $L^{-1}\{f(s)\} = F(t)$, determine $L^{-1}\left\{\dfrac{f(s)}{\sinh cs}\right\}$

Solution

Expand $(\sinh cs)$ in its exponential terms as follows

$$\frac{f(s)}{\sinh cs} = \frac{2f(s)}{e^{cs} - e^{-cs}} \qquad (151)$$

Multiply the numerator and denominator by e^{-cs}, we get

$$\frac{2f(s)}{e^{cs} - e^{-cs}} = \frac{2e^{-cs}f(s)}{1 - e^{-2cs}}$$

Then, use the series expansion of

$$\frac{1}{1 - e^{-2cs}} = \sum_{n=0}^{\infty} e^{-2ncs} \qquad (151\text{-}1)$$

Therefore, (151) becomes

$$\frac{f(s)}{\sinh cs} = 2\sum_{n=0}^{\infty} f(s)e^{-cs}e^{-2ncs} \qquad (151\text{-}2)$$

The inverse Laplace transform of (151-2) is the use of **second translation** or **shifting property** obtained in equation (57-6). Therefore,

$$L^{-1}\left\{\frac{f(s)}{\sinh cs}\right\} = 2\sum_{n=0}^{\infty} F(t - 2nc - c) \ \ U(t - 2nc - c) \qquad (151\text{-}3)$$

Example 85

Solve the differential equation

$$x''(t) + x(t) = F(t) \qquad (152)$$

Given the following initial and boundary conditions

$$x(0) \quad = x'(0) = 0$$

$$
\begin{aligned}
F(t) \quad &= 4 && 0 \le t \le 2 \\
&= t + 2 && t > 2
\end{aligned}
\qquad (152\text{-}1)
$$

Solution

(i) The Laplace transform of the F(t), on the **Right Hand Side** of (152), and given in equation (152-1) is

$$f(s) = \int_0^2 4e^{-st}dt + \int_2^\infty (t+2)e^{-st}dt \qquad (152\text{-}2)$$

The integration in (152-2) is performed in few steps. The Right Hand Side integral is arranged to perform integration-by-parts, in addition to integrating the simple exponential function on the right side of (152-2). Therefore,

$$f(s) = 4\left[\frac{e^{-st}}{-s}\right]_0^2 + \int_2^\infty (t+2)\frac{de^{-st}}{-s}$$

$$= 4\left[\frac{e^{-2s}-1}{-s}\right] + \left(-\frac{1}{s}\right)\left([(t+2)e^{-st}]_2^\infty - \int_2^\infty e^{-st}d(t+2)\right)$$

$$= 4\left[\frac{e^{-2s}-1}{-s}\right] - \frac{1}{s}\left(0 - 4e^{-2s} - \left[\frac{e^{-st}}{-s}\right]_2^\infty\right)$$

$$= 4\left[\frac{e^{-2s}-1}{-s}\right] - \frac{1}{s}\left(-4e^{-2s} + \left[\frac{0-e^{-2s}}{s}\right]\right)$$

$$= 4\left[\frac{e^{-2s}-1}{-s}\right] + \frac{e^{-2s}}{s}\left(\frac{4s+1}{s}\right) = \frac{4}{s} + \frac{e^{-2s}}{s^2} \qquad (152\text{-}3)$$

(ii) The Laplace transform of the **Left Hand Side** of (152) is obtained by the property of derivatives obtained in equations (24-11) and (26-1), in the forms

$$L\{F'(t)\} = s\, f(s) - F(0) .$$

$$L\{F^{(n)}(t)\} = s^n\, f(s) - s^{n-1}\, F(0) - s^{n-2}\, F'(0) - \ldots F^{(n-1)}(0)$$

Therefore,

$$L\{x''(t) + x(t)\} = s^2\, x(s) - s\, x(0) - x'(0) + x(s) \qquad (152\text{-}4)$$

Using the boundary condition given, x(0) = x'(0) = 0, (152-4) becomes

$$L\{x''(t) + x(t)\} = (s^2 + 1)x(s) \tag{152-5}$$

Thus, the Laplace transform of (152) is obtained by the used of R.H.S. in (152-3) and L.H.S. in (152-5) to get

$$(s^2 + 1)x(s) = \frac{4}{s} + \frac{e^{-2s}}{s^2} \tag{152-6}$$

Hence,

$$x(s) = \frac{4}{s(s^2 + 1)} + \frac{e^{-2s}}{s^2(s^2 + 1)}$$

$$= \frac{A}{s} + \frac{Bs + C}{s^2 + 1} + \left(\frac{D}{s^2} + \frac{E}{(s^2 + 1)}\right)e^{-2s} \tag{152-7}$$

Where the constants A, B, C, D, and E are determined as before by equating the coefficients of equal powers of s. Thus,

$A(s^2 + 1) + (Bs + C)s = 4$
$A + B = 0$
$A = 4$
$C = 0$
$B = -A = -4$

$D(s^2 + 1) + E s^2 = 1$
$D + E = 0$
$D = 1$
$E = -D = -1$

Therefore, equation (152-7) becomes

$$x(s) = \frac{4}{s} - \frac{4s}{s^2 + 1} + \left(\frac{1}{s^2} - \frac{1}{(s^2 + 1)}\right)e^{-2s} \tag{152-8}$$

The inverse Laplace transform of (152-8) can be obtained visually as follows:

a. The first term in (1/s) gives a constant   = 4
b. The second term in $(s / (s^2+1)$ gives the cosine term   = -4 cos (t)
c. The third term (e^{-2s} / s^2) gives the linear term shifted by 2:   = (t-2).U(t-2)
d. The fourth term $(e^{-2s} / (s^2+1)$ gives the sine term shifted by 2: - sin(t-2).U(t-2)

Therefore,

$$L^{-1}\left\{\frac{4}{s}-\frac{4s}{s^2+1}+\left(\frac{1}{s^2}-\frac{1}{(s^2+1)}\right)e^{-2s}\right\}=4-4\cos t+[(t-2)-\sin(t-2)].U(t-2) \qquad (152\text{-}9)$$

Example 86

Find $\qquad L^{-1}\left\{\dfrac{1}{s^{\frac{1}{2}}+1}\right\}$ $\qquad\qquad\qquad\qquad (153)$

Solution

The object function can be expressed in its partial fractions as follows

$$\frac{1}{s^{\frac{1}{2}}+1}=\frac{1}{s^{\frac{1}{2}}+1}\left(\frac{s^{\frac{1}{2}}-1}{s^{\frac{1}{2}}-1}\right)=\frac{s^{\frac{1}{2}}-1}{s-1}$$

$$=\frac{s^{\frac{1}{2}}}{s-1}-\frac{1}{s-1}=\left(\frac{s^{\frac{1}{2}}}{s^{\frac{1}{2}}}\right)\frac{s^{\frac{1}{2}}}{s-1}-\frac{1}{s-1}=\left(\frac{1}{s^{\frac{1}{2}}}\right)\frac{s}{s-1}-\frac{1}{s-1}$$

$$=\frac{1}{s^{\frac{1}{2}}(s-1)}+\frac{1}{s^{\frac{1}{2}}}-\frac{1}{s-1} \qquad\qquad\qquad (153\text{-}1)$$

From equation (115-6), we proved that

$$L^{-1}\left\{\frac{1}{(s-1)\sqrt{s}}\right\}=e^t.\mathrm{erf}\left(\sqrt{t}\right)$$

We also proved in (14) that

$$L^{-1}\left\{\sqrt{\frac{\pi}{s}}\right\}=t^{-\frac{1}{2}}$$

Therefore, inverse Laplace transform of (153-1) is

$$L^{-1}\left\{\frac{1}{s^{\frac{1}{2}}+1}\right\}=L^{-1}\left\{\frac{1}{s^{\frac{1}{2}}(s-1)}+\frac{1}{s^{\frac{1}{2}}}-\frac{1}{s-1}\right\}-e^t.\mathrm{erf}(\sqrt{t})+\frac{1}{\sqrt{\pi t}}-e^t$$

$$=\frac{1}{\sqrt{\pi t}}-e^t\left[1-\mathrm{erf}(\sqrt{t})\right]=\frac{1}{\sqrt{\pi t}}-e^t\mathrm{erfc}(\sqrt{t}) \qquad\qquad (153\text{-}2)$$

Example 87

$$\int_{-\infty}^{\infty}\frac{x\sin tx}{a^2+x^2}dx=\pi e^{-at} \qquad a>0 \text{ and } \qquad t>0 \qquad\qquad (154)$$

By **differentiating under the sign of integration** w.r.t. a, deduce the value of

$$\int_{-\infty}^{\infty} \frac{x \sin x}{\left(a^2 + x^2\right)^2}\, dx \tag{154-1}$$

Solution

The integration in (154) is symmetric around the origin of the x axis and can be written as

$$\int_{-\infty}^{\infty} \frac{x \sin tx}{a^2 + x^2}\, dx = 2\int_{0}^{\infty} \frac{x \sin tx}{a^2 + x^2}\, dx \tag{154-2}$$

The Laplace transform of (154-2) is

$$L\left\{\int_{0}^{\infty} \frac{x \sin tx}{a^2 + x^2}\, dx\right\} = \int_{0}^{\infty} e^{-st}\left\{\int_{0}^{\infty} \frac{x \sin tx}{a^2 + x^2}\, dx\right\} dt$$

$$= \int_{0}^{\infty} \frac{x}{a^2 + x^2}\left\{\int_{0}^{\infty} e^{-st} \sin tx \;\; dt\right\} dx$$

$$= \int_{0}^{\infty} \frac{x}{a^2 + x^2} \cdot \frac{x}{s^2 + x^2}\, dx = \int_{0}^{\infty} \frac{x^2}{\left(a^2 + x^2\right)\left(s^2 + x^2\right)}\, dx$$

$$= \int_{0}^{\infty} \left[\frac{A}{\left(s^2 + x^2\right)} + \frac{B}{\left(a^2 + x^2\right)}\right] dx \tag{154-3}$$

Where the constants A and B are obtained as before by equating the coefficients of equal powers of x^2, as follows

$A(a^2 + x^2) + B(s^2 + x^2) = x^2$

$A + B = 1$
$A\,a^2 + B\,s^2 = 0$
$A\,a^2 + B\,a^2 = a^2$
$B = a^2/(a^2 - s^2)$
$A = 1 - a^2/(a^2 - s^2) = -s^2/(a^2 - s^2)$

$$L\left\{\int_{0}^{\infty} \frac{x \sin tx}{a^2 + x^2}\, dx\right\} = \frac{1}{s^2 - a^2}\int_{0}^{\infty} \left[\frac{s^2}{\left(s^2 + x^2\right)} - \frac{a^2}{\left(a^2 + x^2\right)}\right] dx$$

$$= \frac{1}{s^2 - a^2}\left[s \; \tan^{-1}\left(\frac{x}{s}\right) - a \; \tan^{-1}\left(\frac{x}{a}\right)\right]_{0}^{\infty}$$

$$= \frac{1}{s^2 - a^2} \cdot \frac{\pi}{2} (s - a) = \frac{\pi}{2(s + a)} \qquad (154\text{-}4)$$

The inverse Laplace transform of (154-4) is

$$L^{-1}\left\{ L\left\{ \int_0^\infty \frac{x \sin tx}{a^2 + x^2} dx \right\}\right\} = \int_0^\infty \frac{x \sin tx}{a^2 + x^2} dx = L^{-1}\left\{ \frac{\pi}{2(s+a)} \right\} = \frac{\pi}{2} e^{-a} \qquad (154\text{-}5)$$

Therefore,

$$\int_{-\infty}^\infty \frac{x \sin tx}{a^2 + x^2} dx = \pi e^{-a} \qquad (154\text{-}6)$$

Putting t =1 in (154-6) and differentiating under the sign of integration with respect to a, we get

$$\frac{d}{da}\left(\int_{-\infty}^\infty \frac{x \sin tx}{a^2 + x^2} dx \right) = \frac{d}{da}\left(\pi e^{-a} \right)$$

$$-2a \int_{-\infty}^\infty \frac{x \sin tx}{a^2 + x^2} dx = -\pi e^{-a}$$

$$\int_{-\infty}^\infty \frac{x \sin tx}{a^2 + x^2} dx = \frac{\pi}{2a} e^{-a} \qquad (154\text{-}7)$$

Example 88

Solve, using an **inversion integral**, the differential equation

$$(D^2 - 2D + 2)(D^2 + 2D - 3) x = 0 \qquad\qquad t > 0 \qquad (155)$$

Given that

$$\begin{aligned} x(0) &= 1 \\ x'(0) &= 0 \\ x''(0) &= 6 \\ x'''(0) &= -14 \end{aligned}$$

Solution

Equation (155) may be written
$$(D^4 - 5 D^2 + 10 D - 6) x = 0 \qquad (155\text{-}1)$$

The Laplace transform of (155-1) is performed with the aid of equation (26-1), in the forms

$$L\left\{ F^{(n)}(t) \right\} = s^n f(s) - s^{n-1} F(0) - s^{n-2} F'(0) - ...F^{(n-1)}(0)$$

Thus, from (26-1), we write (155-1) as follows

$$[\, s^4 x(s) - s^3 x(0) - s^2 x'(0) - s\, x''\,(0) - x'''\,(0)\,]$$
$$-\,5\quad [\, s^2 x(s) - s\, x(0) - x'(0)\,]$$
$$+\,10\quad [\, s\, x(s) - x(0)\,]\ -6\, x(s) \qquad\qquad = 0 \qquad\qquad (155\text{-}2)$$

Arranging the terms in (155-2) gives

$$s^4 x(s) - 5\, s^2 x(s) + 10\, s\, x(s)\ -6\, x(s)\qquad = s^3 + s\, -4$$
$$(s^4 - 5\, s^2 + 10\, s\ -6)\, x(s)\qquad\qquad = s^3 + s\, -4 \qquad\qquad (155\text{-}2.1)$$

******* Auxiliary Proof *******

In order to put the above polynomial in the form of multiplied parenthesis for the sake of finding partial fractions, we assume the following arrangement

$$s^4 - 5\, s^2 + 10\, s\, -6 = (A\, s^2 + B\, s + C)\,(D\, s^2 + E\, s + F) \qquad\qquad (155\text{-}2.2)$$

Where the five constants A through F are determined as follows

$A = D = 1$	Coefficients of 4th power in s
$B + E = 0$	Coefficients of 3rd power in s
$F + C + EB = -5$	Coefficients of 2nd power in s
$B(F - C) = 10$	Coefficients of 1st power in s
$CF = -6$	Coefficients of zeroth power in s

Hence, let $C = 2$ and $F = -3$

Therefore, the from the Coefficients of 2nd power in s, we get

$-3 + 2 + 5 = B^2 \rightarrow B = \pm\, 2$

Where, $B = -\, E$, from the Coefficients of 3rd power in s.
The sign of B is determined by the Coefficients of 1st power in s as follows.

$-2(\, -3\text{-}2) = -10$

Therefore,
$A = 1$
$B = -2$
$C = 2$
$D = 1$
$E = 2$
$F = -3$

******* End of Auxiliary Proof *******

Thus, equation (155-2.1) becomes

$$(s^2 - 2s + 2)(s^2 + 2s - 3)x(s) = s^3 + s - 4 \qquad (155\text{-}3)$$

$$x(s) = \frac{s^3 + s - 4}{(s - 1)(s + 3)(s^2 - 2s + 2)} \qquad (155\text{-}4)$$

The inverse Laplace transform of (155-4) is obtained by the **Cauchy's Theorem of residues** as follows

$$x(s) = \text{sum of residues of } \frac{\left(s^3 + s - 4\right)e^{st}}{(s - 1)(s + 3)(s - 1 - i)(s - 1 + i)} \text{ at its poles} \qquad (155\text{-}5)$$

****** Determination of Residues ********

Residue at s = 1 + i

$$\frac{\left((1 + i)^3 + 1 + i - 4\right)e^{(i+1)t}}{(1 + i - 1)(1 + i + 3)(i + 1 - 1 + i)} = \frac{(2i - 2 + 1 + i - 4)e^{(i+1)t}}{(2i)i(4 + i)}$$

$$= \frac{(3i - 5)e^{(i+1)t}(4 - i)}{-2(4 + i)(4 - i)} = \frac{(3i - 5)e^{(i+1)t}(4 - i)}{-2(4 + i)(4 - i)} = \frac{e^{(i+1)t}(12i + 3 - 20 + 5i)}{-2(16 + 1)}$$

$$= \frac{17e^{(i+1)t}(i - 1)}{-34} = \frac{e^{(i+1)t}(i - 1)}{-2} = \frac{1}{2}e^{(i+1)t}(1 - i) \qquad (155\text{-}5.1)$$

Residue at s = 1 - i

$$\frac{\left((1 - i)^3 + 1 - i - 4\right)e^{(i-1)t}}{(1 - i - 1)(1 - i + 3)(i - 1 - 1 - i)} = \frac{(-2 - 2i + 1 - i - 4)e^{(i-1)t}}{-2(-i)i(4 - i)}$$

$$= \frac{(-3i - 5)e^{(i-1)t}(4 + i)}{-2(4 - i)(4 + i)} = \frac{(3i + 5)e^{(i-1)t}(4 + i)}{2(4 + i)(4 - i)} = \frac{e^{(i-1)t}(12i + 5i - 3 + 20)}{2(16 + 1)}$$

$$= \frac{17e^{(i-1)t}(i + 1)}{34} = \frac{e^{(i-1)t}(i + 1)}{2} = \frac{1}{2}e^{(i-1)t}(1 + i) \qquad (155\text{-}5.2)$$

Thus the residues at $s = 1 \pm i$ is $\quad \frac{1}{2}(1 \pm i)e^{(1\pm i)t} \qquad (155\text{-}5.3)$

Residue at s = -3

$$\frac{s^3 + s - 4}{(s-1)(s^2 - 2s + 2)} e^{st} = \frac{-27 - 3 - 4}{(-3-1)(9+6+2)} e^{-3t} = \frac{-34}{(-4)(17)} e^{-3t} = \frac{1}{2} e^{-3t} \qquad (155\text{-}5.4)$$

Residue at s = 1

$$\frac{s^3 + s - 4}{(s+3)(s^2 - 2s + 2)} e^{st} = \frac{1 + 1 - 4}{(1+3)(1-2+2)} e^{t} = \frac{-2}{4} e^{t} = -\frac{1}{2} e^{t} \qquad (155\text{-}5.5)$$

Adding the residues from (155-5.3), (155-5.4), and (155-5.5) we get

$$\begin{aligned}
x(t) &= \frac{1}{2}(1 \pm i) e^{(1 \pm i)t} + \frac{1}{2} e^{-3t} - \frac{1}{2} e^{t} + \\
&= \frac{1}{2} e^{t}\left[(1+i)e^{-it} + (1-i)e^{it}\right] + \frac{1}{2} e^{-3t} - \frac{1}{2} e^{t} \\
&= \frac{1}{2} e^{t}\left[(1+i)(\cos t - i\sin t) + (1-i)(\cos t + i\sin t)\right] + \frac{1}{2} e^{-3t} - \frac{1}{2} e^{t} \\
&= \frac{1}{2} e^{t}(2\cos t + 2\sin t) + \frac{1}{2} e^{-3t} - \frac{1}{2} e^{t} \qquad (155\text{-}6)
\end{aligned}$$

Example 89

Prove that $\qquad \int\limits_{0}^{\infty} J_o(t)dt = 1 \qquad\qquad\qquad (156)$

Solution

We have shown in equation (85) that

$$L\{J_o(t)\} = \frac{1}{\sqrt{(s^2 + 1)}}, \qquad s > 0$$

Therefore

$$\int\limits_{0}^{\infty} e^{-st} J_o(t)dt = \frac{1}{\sqrt{(s^2 + 1)}} \qquad (156\text{-}1)$$

Let s $\rightarrow$ 0 + we get

$$\int\limits_{0}^{\infty} e^{-0t} J_o(t)dt = \int\limits_{0}^{\infty} J_o(t)dt = \frac{1}{\sqrt{(0+1)}} = 1 \qquad (156\text{-}2)$$

162

Example 90

Find $\qquad$ L { t sinh 3 t } $\qquad$ (157)

Solution

From equation (19), we have

$$L\{\sinh kt\} = \frac{k}{s^2 - k^2}, \qquad s > |k| \tag{157-1}$$

And, from equation (24-11), we have

$$L\{F'(t)\} = s\ f(s) - F(0) \tag{157-2}$$

Therefore, from (157-1) and (157-2), we can write

$$L\left\{ t\ \sinh 3t \right\} = -\frac{d}{ds}\left\{ \frac{3}{s^2 - 9} \right\} = \frac{6s}{\left(s^2 - 9\right)^2} \tag{157-3}$$

Example 91

Evaluate $\qquad \int\limits_{0}^{\infty} te^{-2t} \sin t\, dt$ $\qquad$ (158)

Solution

In equation (15-14), we have proven that

$$L\{\sin kt\} = \frac{k}{k^2 + s^2} \tag{158-1}$$

And, from equation (24-11), we have

$$L\{F'(t)\} = s\ f(s) - F(0) \tag{158-2}$$

Therefore, from (158-1) and (158-2), we can write

$$L\left\{ t\ \sin t \right\} = -\frac{d}{ds}\left\{ \frac{1}{s^2 + 1} \right\} = \frac{2s}{\left(s^2 + 1\right)^2} \tag{158-3}$$

Therefore, if s → 2, we get

$$\int_0^\infty e^{-st} \sin t\, dt = L\left\{ t \quad \sin t \right\} = \frac{2s}{\left(s^2+1\right)^2} = \frac{2(2)}{\left((2)^2+1\right)^2} = \frac{4}{25} \qquad (158\text{-}4)$$

Which comprises the evaluation of the integral in equation (158).

Example 92

Find $\qquad$ L { F(t) }, $\qquad$ where $\quad$ F(t) $\quad$ = t $\qquad$ $0 < t < 1$

$\qquad\qquad\qquad\qquad\qquad\qquad\qquad\qquad$ = 0 $\qquad\quad$ $1 < t < 2$

$\qquad\qquad\qquad\qquad\qquad\qquad\qquad$ F(t+2) = F(t) $\qquad\qquad\qquad$ (159)

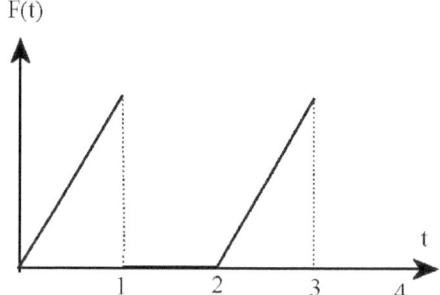

F(t)

1 $\qquad$ 2 $\qquad$ 3 $\qquad$ 4 $\qquad$ t

Solution

In equation (96), we have proven that the periodicity of the function in (159) can be represented as follows

$$f(s) = \frac{\int_0^s e^{-st} F(t)dt}{1 - e^{-as}} \qquad (159\text{-}1)$$

Where, a is the period of the function. Thus, we have

$$L\left\{ F(t) \right\} = \frac{\int_0^2 e^{-st} F(t)dt}{1 - e^{-2s}} \qquad (159\text{-}2)$$

Where, from the given definition of the function, the integral in (159-2) can be written as

$$\frac{\int_0^2 e^{-st} F(t)dt}{1 - e^{-2s}} = \frac{\int_0^1 e^{-st} \, t \, dt + \int_1^2 e^{-st} \, 0 \, dt}{1 - e^{-2s}} = \frac{\int_0^1 e^{-st} \, t \, dt}{1 - e^{-2s}}$$

Integrating by parts, we get

$$= \frac{\int_0^1 t \, d\left(\frac{e^{-st}}{-s}\right)}{1-e^{-2s}} = -\frac{1}{s\left(1-e^{-2s}\right)} \frac{\left[te^{-st}\right]_0^1 - \int_0^1 e^{-st} dt}{1-e^{-2s}}$$

$$= -\frac{1}{s\left(1-e^{-2s}\right)}\left[e^{-s} - \left(\frac{1}{-s}\right)(e^{-s}-1)\right]$$

$$= -\frac{1}{s^2\left(1-e^{-2s}\right)}\left[1 - e^{-s}(s+1)\right]$$

Or

$$L\left\{ F(t) \right\} = -\frac{1}{s^2\left(1-e^{-2s}\right)}\left[1 - e^{-s}(s+1)\right] \tag{159-3}$$

Example 93

Express in terms of the unit step function

$$
\begin{aligned}
F(t) \quad &= \cos 2t \qquad 0 < t < \pi \\
&= \cos 4t \qquad \pi < t < 2\pi \\
&= \cos 6t \qquad t > 2\pi
\end{aligned} \tag{160}
$$

It follows directly from the definition of the **unit step function** that

$$F(t) = \cos 2t + (\cos 4t - \cos 2t)U(t - \pi) + (\cos 6t - \cos 4t)U(t - 2\pi) \tag{161}$$

Example 94

Evaluate

$$\int_0^\infty \frac{e^{-t}\sin t}{t}\,dt \tag{162}$$

Solution

Since

$$L\left\{ \sin t \right\} = \frac{1}{s^2+1}$$

And

$$L\left\{ \frac{\sin t}{t} \right\} = \int_s^\infty \frac{dx}{x^2+1} = \left[\tan^{-1} x \right]_s^\infty$$

$$= \frac{\pi}{2} - \tan^{-1} s \tag{162-1}$$

Therefore, the integral in (162) can be evaluated by letting $s = 1$ in equation (162-1). Thus,

$$\int_0^\infty \frac{e^{-t}\sin t}{t}\,dt = L\left\{ \frac{\sin t}{t} \right\}_{s\to 1} = \frac{\pi}{2} - \tan^{-1} 1 = \frac{\pi}{2} - \frac{\pi}{4} = \frac{\pi}{4} \tag{162-2}$$

Example 95

Prove that $\qquad \displaystyle\int_0^t J_o(u)J_1(t-u)du$ $\qquad\qquad\qquad\qquad\qquad$ (163)

Solution

From (102), the convolution integrals is defined as

$$L^{-1}\{ f_1\,(s)\,f_2\,(s)\,\} = F_1(t)*\,F_2(t) \qquad\qquad (163\text{-}1)$$

Where

$$F_1(t)*F_2(t) = \int_0^t F_1(t-\lambda)\,F_2(\lambda)\,\,d\lambda \qquad\qquad (163\text{-}2)$$

Therefore,

$$\int_0^t J_o(u)J_1(t-u)du = J_o(t)*J_1(t) \qquad\qquad (163\text{-}3)$$

From the property of Laplace transform of derivatives, equation (86), we get

$$s\ \ L\ \left\{J_o(t)\right\} - J_o(0) = L\ \left\{-J_1(t)\right\} \qquad\qquad (163\text{-}4)$$

Therefore, we can write (163) as follows

$$L\left\{\int_0^t J_o(u)J_1(t-u)du\right\} = \frac{1}{\sqrt{s^2+1}}\left(1 - \frac{s}{\sqrt{s^2+1}}\right)$$

$$= \frac{1}{\sqrt{s^2+1}} - \frac{s}{s^2+1} \qquad\qquad (163\text{-}5)$$

The inverse Laplace transform of (165-5) renders the required integral in (163)

$$\int_0^t J_o(u)J_1(t-u)du = L^{-1}\left\{\frac{1}{\sqrt{s^2+1}} - \frac{s}{s^2+1}\right\} = J_o(t) - \cos t = J_o(t) - \cos t \qquad (163\text{-}6)$$

Example 96

Solve the integral equation

$$\int_0^t Y(t-\lambda)Y(\lambda)d\lambda = 8\sin 2t \qquad\qquad (164)$$

Solution

The convolution objects $Y(t)*\ Y(t) = 8\sin 2t$.

Therefore,

$$[y(s)]^2 = \frac{16}{s^2 + 4} \tag{164-1}$$

$$y(s) = \pm \frac{4}{\sqrt{s^2 + 4}} \tag{164-2}$$

Now, $L\{J_o(at)\} = \dfrac{1}{\sqrt{s^2 + a^2}}$ \hfill (164-3)

$$Y(t) = L^{-1}\{y(s)\} = \pm L^{-1}\left\{\frac{4}{\sqrt{s^2 + 4}}\right\} = \pm 4J_o(2t) \tag{164-4}$$

Example 97

Find   L { Log t }   (165)

Solution

We have shown in equation (12) that

$$L\{t^k\} = \frac{\Gamma(k+1)}{s^{k+1}} \qquad \text{where } k > -1 \tag{165-1}$$

i. e.,

$$\int_0^\infty e^{-st}\{t^k\}\,dt = \frac{\Gamma(k+1)}{s^{k+1}} \tag{165-2}$$

Differentiate (165-2) w,r.t.. k

$$\frac{d}{dk}\left(\int_0^\infty e^{-st}\{t^k\}\,dt\right) = \frac{d}{dk}\left(\frac{\Gamma(k+1)}{s^{k+1}}\right) \tag{165-3}$$

The derivative of t^k is obtained as follows
$$x = t^k \tag{165-3.1}$$
$$\log x = k \log t$$
$$\frac{d}{dk}\log x = \log t$$
$$\frac{dx}{dk}\left(\frac{1}{x}\right) = \log t$$
$$\frac{dx}{dk} = x \log t = t^k \log t \tag{165-3.2}$$

Therefore, (165-3) becomes

$$\int_0^\infty e^{-st}\{t^k \log t\}\, dt = \frac{\Gamma''(k+1) - \Gamma'(k+1)\log(s)}{s^{k+1}} \qquad (165\text{-}4)$$

Put $k = 0$

$$\int_0^\infty e^{-st}\log t\, dt = \frac{\Gamma'(1) - \Gamma(1)\log(s)}{s} = \frac{\Gamma'(1) - \log(s)}{s} \qquad (165\text{-}5)$$

Example 98

Solve the integral equation

$$Y(t) = a \sin 8t + 6 \int_0^t Y(\lambda) \sin 8(t-\lambda)\, d\lambda \qquad (166)$$

Solution

The integral in (166) take the form of convolution integral as follows

$$Y(t) = a \sin 8t + 6\; Y(t) * \sin 8t \qquad (166\text{-}1)$$

The Laplace transform of (166-1) is

$$y(s) = \frac{8a}{s^2 + 64} + 6 y(s) \frac{8}{s^2 + 64}$$

$$y(s)\left[1 - \frac{48}{s^2 + 64}\right] = y(s)\left[\frac{s^2 + 16}{s^2 + 64}\right] = \frac{8a}{s^2 + 64}$$

$$y(s) = \frac{8a}{s^2 + 16} \qquad (166\text{-}2)$$

The inverse Laplace transform of (166-2) is

$$Y(t) = L^{-1}\left\{\frac{8a}{s^2 + 16}\right\} = 2a \sin 4t \qquad (166\text{-}3)$$

Example 99

Evaluate $\qquad \int_0^\infty \int_0^t \frac{e^{-t}\sin u}{u}\, dt\, du \qquad (167)$

Solution

Interchanging the order of integration in (167), we get

$$\int_0^\infty e^{-t}\left\{\int_0^t \frac{\sin u}{u}\, du\right\} dt \qquad (167\text{-}1)$$

First, we start with the elementary Laplace transform

$$L\{\sin t\} = \frac{1}{s^2 + 1} \qquad (167\text{-}2)$$

Second, we use the integration property of the Laplace transform upon division of the object function as follows

$$L\left\{\frac{\sin t}{t}\right\} = \int_s^\infty \frac{dx}{x^2 + 1} = \left[\tan^{-1} x\right]_s^\infty$$

$$= \frac{\pi}{2} - \tan^{-1} s \qquad (167\text{-}3)$$

$$= \tan^{-1}\frac{1}{s} \qquad (167\text{-}4)$$

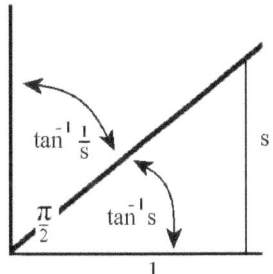

Third, the Laplace transform of (167-4) is

$$\int_0^\infty \int_0^t \frac{e^{-t}\sin u}{u}\, dt\; du = L\left\{L\left\{\frac{\sin t}{t}\right\}\right\} = L\left\{\tan^{-1}\frac{1}{s}\right\} = \frac{1}{s}\tan^{-1}\frac{1}{s} \qquad (167\text{-}5)$$

Where the division by s resulted from the integration of the object function over t.

Let s → 1, equation (167-5) gives

$$\int_0^\infty \int_0^t \frac{e^{-t}\sin u}{u}\, dt\; du = \frac{1}{1}\tan^{-1}\frac{1}{1} = \frac{\pi}{4} \qquad (167\text{-}6)$$

Example 100

Evaluate $\displaystyle\int_0^\infty \frac{e^{-2t} - e^{-4t}}{t}\, dt$ (168)

Solution

First, start with elementary functions as follows

$$L\left\{ e^{-2t} \right\} = \frac{1}{s+2} \tag{168-1}$$

$$L\left\{ e^{-4t} \right\} = \frac{1}{s+4} \tag{168-2}$$

$$L\left\{ e^{-2t} - e^{-4t} \right\} = \frac{1}{s+2} - \frac{1}{s+4} \tag{168-3}$$

Second, implement the division by t, of the object function under the Laplace brackets that implies integration of the product function as follows

$$L\left\{ \frac{e^{-2t} - e^{-4t}}{t} \right\} = \int_s^\infty \left[\frac{1}{s+2} - \frac{1}{s+4} \right] dx = \left[\log \frac{x+2}{x+4} \right]_s^\infty = \log \frac{s+4}{s+2} \tag{168-4}$$

i.e.,

$$\int_0^\infty e^{-st}\, \frac{e^{-2t} - e^{-4t}}{t}\, dt = \log \frac{s+4}{s+2} \tag{168-5}$$

In order to arrive at the required solution, let s → 0

$$\int_0^\infty e^{-st}\, \frac{e^{-2t} - e^{-4t}}{t}\, dt = \log \frac{0+4}{0+2} = \log 2 \tag{168-6}$$

Example 101

Evaluate $\displaystyle\int_0^\infty e^{-t} J_0(t)\, dt$ (169)

Solution

We have seen that

$$L\left\{ J_0(t) \right\} = \int_0^\infty e^{-st}\, J_0(t)\ dt = \frac{1}{\sqrt{s^2 + 1}} \tag{169-1}$$

Thus, the required solution is attained by letting s $\to$ 1. Therefore, from (169-1), we get

$$\int_0^\infty e^{-st} \ J_0(t) \ dt = \frac{1}{\sqrt{1+1}} = \frac{1}{\sqrt{2}} \tag{169-2}$$

Example 102

Evaluate $\quad L^{-1}\left\{\dfrac{1}{s^2(s^2+1)}\right\}$ $\hfill$ (170)

(i)　by resolving into partial fractions
(ii)　by convolution

(iii)　from the Laplace transform of elementary functions $L^{-1}\left\{\dfrac{1}{(s^2+1)}\right\}$

(iv)　by using inversion integral.

Solution

(i) Partial Fraction

$$L^{-1}\left\{\frac{1}{s^2(s^2+1)}\right\} = L^{-1}\left\{\frac{1}{s^2} - \frac{1}{s^2+1}\right\} = t - \sin t \tag{170-1}$$

(ii) Convolution integral

$$L^{-1}\left\{\frac{1}{s^2(s^2+1)}\right\} = t * \sin t = \int_0^t (t-\lambda)\sin \lambda d\lambda$$

$$= -\int_0^t (t-\lambda) \ d\cos \lambda = [-(t-\lambda)\cos \lambda - \sin \lambda]_0^t = -\sin t + t \tag{170-2}$$

(iii) Elementary Function

$$L^{-1}\left\{\frac{1}{(s^2+1)}\right\} = \sin t \tag{170-3.1}$$

Using the property of division of the transform by and integrating the object function, we get

$$L^{-1}\left\{\frac{1}{s(s^2+1)}\right\} = \int_0^t \sin \tau d\tau = [-\cos \tau]_0^t = 1 - \cos t \tag{170-3.2}$$

Repeating the same process one more time, we get

$$L^{-1}\left\{\frac{1}{s^2(s^2+1)}\right\} = \int_0^t (1-\cos\tau)d\tau = [\tau - \sin\tau]_0^t = t - \sin t \qquad (170\text{-}3.3)$$

(iv) by using inversion integral.

$$L^{-1}\left\{\frac{1}{s^2(s^2+1)}\right\} = \text{sum of residues of}\left\{\frac{e^{st}}{s^2(s-i)(s+i)}\right\} \text{ at its poles} \qquad (170\text{-}4)$$

Residues at s = 0:

$$\text{residue} = \frac{1}{1!}\left[\frac{d}{ds}\left(\frac{s}{s^2+1}e^{st}\right)\right]_{s=0} = \left[\frac{(s^2+1)te^{st} - 2se^{st}}{(s^2+1)^2}\right]_{s=0} = t \qquad (170\text{-}4.1)$$

Residues at s = ± i

$$\text{residue} = \left[\frac{e^{st}}{s^2(s+i)}\right]_{s=+i} + \left[\frac{e^{st}}{s^2(s-i)}\right]_{s=-i} = \frac{e^{it}}{-2i} + \frac{e^{-it}}{2i} == \frac{e^{-it} - e^{it}}{2i} = -\sin t \qquad (170\text{-}4.2)$$

Therefore, from (170-4.1) and (170-4.2), the sum of residues is

$$L^{-1}\left\{\frac{1}{s^2(s^2+1)}\right\} = t - \sin t \qquad (170.4.3)$$

Example 103

Solve the equation

$$t\ Y''(t) + (t\text{-}1)\ Y'(t) - Y(t) = 0 \qquad (171)$$

Given the initial conditions

$$Y(0) = 5$$
$$Y(\infty) = 0$$

Solution

The Laplace transform of (171) is

$$-\frac{d}{ds}[s^2 y(s) - sY(0) - Y'(0)] - \frac{d}{ds}[sy(s) - Y(0)] - [sy(s) - Y(0)] - y(s) = 0 \qquad (171\text{-}1)$$

Substituting by the given initial conditions in (171-1), we get

$$-\frac{d}{ds}[s^2 y(s) - 5s - A] - \frac{d}{ds}[sy(s) - 5] - [sy(s) - 5] - y(s) = 0$$

Therefore,

$$-s^2 \frac{dy(s)}{ds} - 2sy(s) + 5 - s\frac{dy(s)}{ds} - y(s) - sy(s) + 5 - y(s) = 0$$

$$-(s^2 + s)\frac{dy(s)}{ds} - 3sy(s) - 2y(s) + 10 = 0$$

$$\frac{dy(s)}{ds} + \frac{3s + 2}{s^2 + s}y(s) = \frac{10}{s^2 + s} \tag{171-2}$$

Equation (171-2) is a **linear differential equation of the first order**.

In order to determine the I.F. (Integrating Factor), equation (171-2) can be written by expanding the fraction multiplied by y(s) as follows.

$$\frac{dy(s)}{ds} + \left(\frac{2}{s} + \frac{1}{s+1}\right)y(s) = \frac{10}{s^2 + s} \tag{171-3}$$

The I.F. is defined as follows

$$\text{I.F.} = \exp\left(\int \left[\frac{2}{s} + \frac{1}{s+1}\right]ds\right) = \exp(2\log s + \log(s+1))$$
$$= \exp\left(\log\left[s^2 (s+1)\right]\right) = s^2(s+1) \tag{171-4}$$

Hence, equation (171-3) can be written as

$$y(s).(\text{I.F.}) = \int \frac{10}{s^2 + s}.(\text{I.F.})ds + C \tag{171-5}$$

$$y(s).s^2(s+1) = \int \frac{10}{s^2 + s}s^2(s+1).ds + C = 10\int s.ds + C$$
$$= 5s^2 + C \tag{171-6}$$

Therefore,

$$y(s) = \frac{5}{s+1} + \frac{C}{s^2(s+1)} \tag{171-7}$$

By expressing it in the partial fraction form, equation (171-7) becomes

$$y(s) = \frac{5}{s+1} + C\left(\frac{1}{s^2} - \frac{1}{s} + \frac{1}{s+1}\right) \tag{171-8}$$

The inverse Laplace transform of (171-8) is

173

$$Y(t) = L^{-1}\left\{\frac{5}{s+1} + C\left(\frac{1}{s^2} - \frac{1}{s} + \frac{1}{s+1}\right)\right\} = 5e^{-t} + C(t - 1 + e^{-t}) \qquad (171\text{-}9)$$

C is obtained from the initial conditions by substituting $Y(\infty) = 0$, then $C = 0$.

$$Y(t) = 5e^{-t} \qquad (171\text{-}10)$$

Example 104

Solve the differential equation

$$\frac{d^2y}{dx^2} - a^2y = \delta(x - b) \qquad 0 < x < 1 \qquad 0 < b < 1 \qquad (172)$$

Given that

$$y = 0 \text{ when } x = 0 \text{ and } x = 1$$

Or

$$y(0) = y(1) = 0$$

Solution

From equation (70-1), we have seen that

$$L\left\{\delta(t - T)\right\} = e^{-Ts}$$

Therefore, Laplace transform of the differential equation (172) is

$$s^2y(s) - sy(0) - y'(0) - a^2y(s) = e^{-bs} \qquad (172\text{-}1)$$

Put $\qquad y'(0) = A$

Therefore, (172-1) becomes

$$(s^2 - a^2)y(s) = A + e^{-bs} \qquad (172\text{-}2)$$

$$y(s) = \frac{A}{s^2 - a^2} + \frac{e^{-bs}}{s^2 - a^2} \qquad (172\text{-}3)$$

The inverse Laplace transform of (172-3) contains the (sinh ax) and the step unit function (U(x-b)) as follows

$$y(x) = L^{-1}\left\{\frac{A}{s^2 - a^2} + \frac{e^{-bs}}{s^2 - a^2}\right\} = \frac{A}{a}\sinh ax + \frac{1}{a}\sinh a(x - b)U(x - b) \qquad (172\text{-}4)$$

174

The constant A is determined by substituting the given boundary conditions $y(0) = y(1) = 0$, as follows. At $y(1) = 0$,

$$0 = \frac{A}{a}\sinh a + \frac{1}{a}\sinh a(1-b)$$

Therefore, the constant A is determined by the expression

$$A = -\frac{\sinh a(1-b)}{\sinh a} \tag{172-5}$$

Substituting by A from (172-5) in (172-4), we get

$$y(x) = -\frac{1}{a}\cdot\frac{\sinh a(1-b)}{\sinh a}\sinh ax + \frac{1}{a}\sinh a(x-b)U(x-b) \tag{172-6}$$

Equation (172-6) implies that

$$
\begin{array}{lll}
U(x\text{-}b) & = 0 & \text{when } 0 < x < b \\
& = 1 & \text{when } b < x < 1
\end{array}
$$

******* when $0 < x < b$ → U(x-b) = 0 *********

$$y(x) = -\frac{1}{a}\cdot\frac{\sinh a(1-b)}{\sinh a}\sinh ax \tag{172-7}$$

******* when $b < x < 1$ → U(x-b) = 1 *******

$$y(x) = -\frac{1}{a}\cdot\frac{\sinh a(1-b)}{\sinh a}\sinh ax + \frac{1}{a}\sinh a(x-b)$$

$$= \frac{1}{a\sinh a}\big(\sinh a(x-b)\sinh a - \sinh a(1-b)\sinh ax\big)$$

$$= \frac{1}{a\sinh a}\cdot\frac{1}{4}\Big[\big(e^{a(x-b)} - e^{-a(x-b)}\big)\big(e^{a} - e^{-a}\big) - \big(e^{a(1-b)} - e^{-a(1-b)}\big)\big(e^{ax} - e^{-ax}\big)\Big]$$

$$= \frac{1}{a\sinh a}\cdot\frac{1}{4}\Big[e^{a(x-b+1)} - e^{-a(x-b-1)} - e^{a(x-b-1)} + e^{-a(x-b+1)}\big) - \big(e^{a(1-b+x)} - e^{-a(1-b-x)} - e^{a(1-b-x)} + e^{-a(1-b+x)}\big)\Big]$$

$$-\frac{1}{a\sinh a}\cdot\frac{1}{4}\Big[\big(-e^{-a(x-b-1)} - e^{a(x-b-1)}\big) - \big(-e^{-a(1-b-x)} - e^{a(1-b-x)}\big)\Big]$$

$$= \frac{1}{a\sinh a}\cdot\frac{1}{4}\Big[e^{ab}\big(e^{a(x-1)} - e^{-a(x-1)}\big) - e^{-ab}\big(e^{a(x-1)} - e^{a(1-x)}\big)\Big]$$

$$= \frac{1}{a\sinh a}\cdot\frac{1}{4}\Big[\big(e^{ab} - e^{-ab}\big)\big(e^{a(x-1)} - e^{-a(x-1)}\big)\Big]$$

$$= \frac{1}{a\sinh a}\cdot\sinh ab\sinh a(x-1) \tag{172-8}$$

175

2.26. Exercises on Laplace Transforms

(1) Find

$$\text{(i) } L^{-1}\left\{\frac{5e^{-3s}}{s} - \frac{e^{-s}}{s}\right\}$$

$$\text{(ii) } L^{-1}\left\{\frac{e^{-4s}}{(s+2)^3}\right\}$$

(2) If F(t) is to be continuous for $t \geq 0$ and

$$F(t) = L^{-1}\left\{\frac{(1-e^{-2s})(1-3e^{-2s})}{s^2}\right\}$$

Then evaluate F(1), F(3), and F(5).

(3) Find the Laplace transform of

$$F(t) = E \sin \omega t \qquad 0 < t < \frac{\pi}{\omega}$$

$$=0 \qquad\qquad t > \frac{\pi}{\omega}$$

(4) Find

(i) $L\{t^2 \sin kt\}$

(ii) $L\{t^2 \cos kt\}$

(5) Find the Laplace transform of the periodic function $\psi(t,c)$, where

$$\psi(t,c) \quad = 1 \qquad\qquad 0 < t < c$$
$$\qquad\qquad =0 \qquad\qquad c < t < 2c$$
$$\psi(t + 2c) = \psi(t, c),$$

(6) Evaluate $L\left\{\dfrac{\sin kt}{t}\right\}$

(7) Evaluate $L\left\{\dfrac{\sinh kt}{t}\right\}$

(8) Evaluate $L\left\{\dfrac{1-\cosh kt}{t}\right\}$

(9) Evaluate $L^{-1}\left\{\dfrac{1}{s^3(1-e^{-2s})}\right\}$ and compute F(5).

(10) If $\varphi(t) = L^{-1}\left\{\dfrac{3}{s^4 \sinh 3s}\right\}$ and compute $\varphi(10)$.

(11) Prove that

$$L^{-1}\{f(s)\tanh cs\} = F(t) + 2\sum_{n=1}^{\infty} (-1)^n F(t - 2nc)(U(t - 2nc) \quad c > 0, \; s > 0$$

(12) Prove that

$$L^{-1}\left\{\dfrac{f(s)}{\cosh cs}\right\} = 2\sum_{n=0}^{\infty} (-1)^n F(t - 2nc - c)(U(t - 2nc - c) \quad c > 0, \; s > 0$$

(13) Compute x(t) and x(4) for the function x(t) which satisfies the boundary-value problem

$$x''(t) + 2x'(t) + x(t) = 2 + (t - 3)\, U(t - 3)$$

given that

$$x(0) = 2 \text{ and } x'(0) = 1$$

(14) Solve the differential equation

$$x''(t) + 2x'(t) + x(t) = F(t)$$
Given $\quad x(0) = x'(0) = 0$

(15) Solve the differential equation

$$y''(t) + 4y'(t) + 13y(t) = F(t)$$
Given $\quad y(0) = y'(0) = 0$

Solve the differential equations

(16) $\quad x''(t) + 6x'(t) + 9x(t) = F(t)$

Given $\quad x(0) = A$
$\quad\quad\quad x'(0) = B$

(17) $\quad F(t) = 1 + 2\displaystyle\int_0^t e^{-2\lambda}F(t - \lambda)d\lambda$

(18) $\quad F(t) = 4t^2 - \displaystyle\int_0^t F(t - \lambda)d\lambda$

(19) $F(t) = 8t^2 - 3\int\limits_0^t F(\lambda)\sin t - \lambda)d\lambda$

Solve the following differential equations and check by using an inversion integral:

(20) $(D^3+1)\,x = 1$

        Given that

        $x(0) = x'(0) = x''(0) = 0$

(21) $(D+1)\,(D+2)\,(D+3)\,x = 1 + t + t^2$

        Given that

        $x(0) = x'(0) = x''(0) = 0$

(22) $D\,(D - 1)\,x = t^2$

        Given that

        $x(0) = x_o$ and
        $x'(0) = x_1$

(23) $(D^2 - D + 2)\,x = e^t$

        Given that

        $x(0) = x_o$ and
        $x'(0) = x_1$

(24) Indicate which of the following are null functions

(i) $F(t)$ $= 2,$ $t = 7=$
                   $= 0$ Otherwise

(ii) $F(t)$ $= 4$ $1 \le t \le 3$

(25) Given that

$L\left\{ \sin \sqrt{t} \right\} = \dfrac{\sqrt{\pi}}{2s^{\frac{3}{2}}}\,e^{-\frac{1}{4s}}$

Show that

178

$$L\left\{ \frac{\cos \sqrt{t}}{\sqrt{t}} \right\} = \frac{\sqrt{\pi}}{s^{\frac{7}{2}}} e^{-\frac{1}{4s}}$$

(26) If $L\left\{ F(t) \right\} = \frac{e^{-\frac{1}{s}}}{s}$, find $L\left\{ e^{-t}F(4t) \right\}$

(27) Show that $L\left\{ \int_{0}^{t} \frac{1-e^{-\tau}}{\tau} d\tau \right\} = \frac{1}{s}\log\left(1+\frac{1}{s}\right)$

(28) Find $L\{F(t)\}$ where

$$\begin{aligned} F(t) &= t^2 \qquad 0 < t < 2 \\ F(t+2) &= F(t) \end{aligned}$$

(29) Express in terms of the unit step function

$$\begin{aligned} F(t) &= t^3 \qquad 0 < t < 3 \\ &= 5t^2 \qquad\ \ t > 3 \end{aligned}$$

(30) Prove that $\displaystyle\int_{0}^{\infty} t^3 e^{-\lambda} \sin t\, dt = 0$

(31) Find $L\{F(t)\}$ where

$$\begin{aligned} F(t) &= \cos t \qquad 0 < t < \pi \\ &= \sin t \qquad\ \ t > \pi \end{aligned}$$

(32) Prove that $L\left\{ \sin^5 t \right\} = \dfrac{120}{(s^2+1)(s^2+9)(s^2+25)}$

(33) Prove that $\displaystyle\int_{0}^{\infty} \cos x^2 dx = \frac{1}{2}\sqrt{\frac{\pi}{2}}$

(34) Find by convolution

(i) $L^{-1}\left\{ \dfrac{1}{(s+2)^2(s-2)} \right\}$

179

(ii) $L^{-1}\left\{\dfrac{s^2}{\left(s^2+4\right)^2}\right\}$

(iii) $L^{-1}\left\{\dfrac{1}{(s+1)\left(s^2+1\right)}\right\}$

(35) Using Heaviside's expansion formula, find

$$L^{-1}\left\{\dfrac{1}{(s+1)\left(s^2+1\right)}\right\}$$

(36) Prove that

$$L^{-1}\left\{\dfrac{1}{s}\cos\dfrac{1}{s}\right\}=1-\dfrac{t^2}{(2!)^2}+\dfrac{t^4}{(4!)^2}-\dfrac{t^6}{(6!)^2}+..$$

(37) Solve the equation

$$t\,Y''(t)+2\,Y'(t)+t\,Y(t)=0$$

Given that

$$Y(0+)=1 \text{ and}$$
$$Y(\pi)=0$$

(38) Solve the equation

$$Y''(t)+t\,Y'(t)-Y(t)=0$$

Given that

$$Y(0) \;=0 \text{ and}$$
$$Y'(0)=1$$

(39) Solve the equation

$$tY''(t)+(1-2t)\,Y'(t)-2Y(t)=0$$

Given that

$$Y(0) \;=1 \text{ and}$$
$$Y'(0)=12$$

(40) Solve the integral equation

$$Y(t)=t^2+\int_0^t Y(\lambda)\sin(t-\lambda)d\lambda$$

(41) Solve for Y(t) the second order difference equation

$$Y(t) - (a + b)Y(t - h) + ab\, Y(t - 2h) = F(t)$$

$$h > 0$$

Given that

$$Y(t) = 0 \text{ and}$$
$$F(t) = 0 \text{ when } t < 0 \text{ in the case when}$$

(i) $a \neq b$

(ii) $a = b$

(42) Solve the integral equation

$$Y(t) = 6t + \int_0^t Y(t - \lambda)\sin \lambda\, d\lambda$$

Solve the equations

(43) $$Y''(t) + 2t\, Y'(t) - 4Y(t) = 1$$

Given that

$$Y(0) = Y'(0) = 0$$

(44) $$tY''(t) - (1 + t)\, Y'(t) + 2Y(t) = t - 1$$

Given that

$$Y(0) = 0$$

(45) $$Y'(t) + k^2 \int_0^t Y(\lambda)\cosh k(t - \lambda).d\lambda = 0$$

(46) Prove that

$$L\left\{\frac{e^t - \cos t}{t}\right\} = \frac{1}{2}\log\left\{\frac{s^2 + 1}{(s - 1)^2}\right\}$$

(47) Prove that

$$\int_0^\infty \frac{\cos tx}{x^2 + a^2}.dx = \frac{\pi}{2a^2}e^{-at}, \qquad > 0,\, t \geq 0$$

(48) Prove that

$$\int_0^\infty \frac{1 - \cos 2tx}{x^2}.dx = \pi t$$

(49) Using an inversion integral find the following inverse Laplace Transform

181

$$L^{-1}\left\{\frac{2s+1}{s(s^2+1)}\right\}$$

(50) Prove that $L\left\{t\ J_0(at)\right\} = \dfrac{s}{(s^2+a^2)^{3/2}}$

(51) Find $L\left\{t\ J_1(t)\right\}$

(52) Solve the integral equation

$$Y(t) = \cos t - 3\int_0^t Y(\lambda)\sin(t-\lambda).d\lambda$$

(53) Solve the difference equations

(i) $Y(t) - Y(t-2) = t$, where $Y(t) = 0$ when $t < 0$
(ii) $Y'(t) - Y(t-1) = t$, where $Y(t) = 0$ when $t \leq 0$

(54) Find the Laplace transform of the wave function $f(c,t)$ defined by

$$\begin{aligned}
f(c,t) \quad &= 0 &\quad 0 < t < c \\
&= t - c &\quad c < t < 2c \\
f(t+2c) &= F(t)
\end{aligned}$$

(55) Find inverse Laplace transform $L^{-1}\left\{\dfrac{1}{s^2(s-1)}\right\}$ by the following four methods:

(i) by partial fractions.
(ii) by convolution integral.
(iii) from the elementary inverse Laplace transform $L^{-1}\left\{\dfrac{1}{s-1}\right\}$.
(iv) by an inversion integral.

(56) Solve the differential-integral equation

$$Y(t) + 2\int_0^t Y(\lambda)\cosh(t-\lambda).d\lambda = \sinh t$$

(57) Find inverse Laplace transform $L^{-1}\left\{\dfrac{1}{s(s^2-1)}\right\}$ using an inversion integral.

(58) Solve the differential-integral equation

$$Y'(t) = \sin t + \int_0^t Y(t-\lambda)\cos\lambda.d\lambda$$

Given that $Y(0) = 0$.

(59) Find the Laplace transform of the function $f(c,t)$ defined by

$$
\begin{array}{llll}
f(c,t) & = 1 & & 0 < t < c \\
 & = 0 & & c < t < 2c \\
 & = 1 & & 2c < t < 3c \\
f(c, t + 3c) & = F(c,t)
\end{array}
$$

(60) Solve the differential-integral equation

$$Y'(t) = \sinh t - \int_0^t Y(t - \lambda)\cosh \lambda . d\lambda$$

Given that $Y(0) = 0$.

(61) Solve the difference equation

$$Y(t) - 4Y(t - h) + 4 Y(t - 2h) = t^2$$

$$h > 0$$

Given that

$$Y(t) = 0 \text{ and}$$
$$F(t) = 0 \text{ when } t < 0 \text{ in the case when}$$

(62) Find the Laplace transform of the wave function $f(c,t)$ defined by

$$
\begin{array}{llll}
f(c,t) & = 2 & & 0 < t < c \\
 & = -2 & & c < t < 2c \\
 & = 2 & & 2c < t < 3c
\end{array}
$$

$$f(, t + 3c) = F(c,t)$$

(63) Solve the differential-integral equation

$$Y'(t) + \int_0^t Y(\lambda)\cosh(t - \lambda). d\lambda = 0$$

(64) Find the Laplace transform of

$$
\begin{array}{llll}
F(t) & = 0 & & 0 < t < (a/3) \\
 & = c & & (a/3) < t < (2a/3) \\
 & = 0 & & (2a/3) < t < a
\end{array}
$$

$$F(t + a) = F(t)$$

(65) Evaluate $\int\limits_{0}^{\infty} \dfrac{\cos 2t - \cos 4t}{t}.dt$

(66) Evaluate $\int\limits_{0=\infty}^{\infty} e^{-t}U(t-3)dt$

(67) Find $L\left\{ te^{-3t}J_{o}\left(t\sqrt{2}\right) \right\}$

(68) Using an inversion integral find the following inverse Laplace Transform

$$L^{-1}\left\{ \dfrac{5s-2}{s(s+1)(s11)} \right\}$$

(69) Solve the difference equation

$$Y'(t) + Y(t-1) = t^2$$

Given that
$$Y(t) \;=0 \quad\text{for}\quad t\le 0$$

(70) Using an inversion integral find the following inverse Laplace Transform

$$L^{-1}\left\{ \dfrac{1}{s^2(s^2+16)} \right\}$$

Answers

(1)
(i) $5\,U(t\text{-}3) - U(t\text{-}1)$

(ii) $\dfrac{1}{2}(t-4)^2 e^{-2(t-4)}U(t-4)$

(2) $F(1) = 1,\ F(3) = -1,\ F(5) = -4$

(3) $f(s) = \dfrac{E\omega}{s^2+\omega^2}\left(1 + e^{-\frac{s\pi}{\omega}}\right)$

(4)
(i) $L\left\{t^2 \sin kt\right\} = \dfrac{2k(3s^2 - k^2)}{(s^2 + k^2)^3},\ s>0$

(ii) $L\left\{t^2 \cos kt\right\} = \dfrac{2s(s^2 - 3k^2)}{(s^2 + k^2)^3},\ s>0$

(5) $L\left\{\psi(t,c)\right\} = \dfrac{1}{s(1+e^{-cs})}$

(6) $\qquad L\left\{\dfrac{\sin kt}{t}\right\} = \tan^{-1}\dfrac{k}{s}, \quad s > 0$

(7) $\qquad L\left\{\dfrac{\sinh kt}{t}\right\} = \dfrac{1}{2}\log\dfrac{s+k}{s-k}, \qquad\qquad s > k > 0$

(8) $\qquad L\left\{\dfrac{1-\cosh kt}{t}\right\} = \dfrac{1}{2}\log(-\dfrac{k^2}{s^2}) \qquad s > k > 0$

(9) $\qquad L^{-1}\left\{\dfrac{1}{s^3(1-e^{-2s})}\right\} = F(t) = \dfrac{1}{2}\sum\limits_{n=0}^{\infty} (t-2n)^2(U(t-2n)$

$\qquad F(5) = \dfrac{1}{2}\left[(5-0)^2(U(5-0)+(5-2)^2(U(5-2)+(5-4)^2(U(5-4)\right] = \dfrac{1}{2}[25+9+1] = \dfrac{35}{2}$

$\qquad = 17.5$

(10) $\qquad \varphi(10) = 344$

(11) $\qquad$ Proof steps required

(12) $\qquad$ Proof steps required

(13) $\qquad x(1) = 2 + \dfrac{1}{e}$

$\qquad x(4) = 1 + \dfrac{3}{e} + \dfrac{4}{e^4}$

(14) $\qquad x(t) = \displaystyle\int_0^t \lambda e^{-\lambda}F(t-\lambda)d\lambda$

(15) $\qquad y(t) = \dfrac{1}{3}\displaystyle\int_0^t e^{-2\lambda}\sin 3\lambda F(t-\lambda)d\lambda$

(16) $\qquad y(t) = e^{-3\lambda}[A+(B+3A)t] + \displaystyle\int_0^t \lambda e^{-3\lambda}F(t-\lambda)d\lambda$

(17) $\qquad F(t) = 1 + 2t$

(18) $\qquad F(t) = -1 + 2t + 2t^2 + e^{-2t}$

(19) $\qquad F(t) = 2t + 3 - 3\cos 2t$

(20) $\qquad x = 1 - \dfrac{1}{3}e^{-t} - \dfrac{2}{3}e^{-\frac{1}{2}t}\cos\left(\dfrac{1}{2}t\sqrt{3}\right)$

(21) $\qquad \dfrac{35}{54} - \dfrac{4}{9}t + \dfrac{1}{6}t^2 - e^{-t} + \dfrac{1}{2}e^{-2t} - \dfrac{4}{27}e^{-3t}$

(22) $\qquad (x_o - x_1) + x_1 e^t - 2(1 + t + \dfrac{1}{2!}t^2 + \dfrac{1}{3!}t^3 - e^t)$

(23) $(x_o - x_1 + 1)e^{2t} + (2x_o - x_1 - 1 - t)e^t$

(24)   (i) is a null function because the area under the dot is zero
         (ii) is not a null function because it contains greater than zero area.

(25)   Proof steps required

(26) $L\left\{ e^{-t}F(4t) \right\} = \dfrac{e^{-\frac{4}{s+1}}}{s+1}$

(27)   Proof steps required

(28) $\dfrac{2 - 2e^{-2s} - 4se^{-2s} - 4s^2e^{-2s}}{s^3(t - e^{-2s})}$

(29) $F(t) = t^3 + (5t^2 - t^3)\,U(t - 3)$

(30)   Proof steps required

(31) $\dfrac{s + (s-1)e^{-\pi s}}{s^2 + 1}$

(32)   Proof steps required

(33)   Proof steps required

(34)

(i) $L^{-1}\left\{ \dfrac{1}{(s+2)^2(s-2)} \right\} = \dfrac{1}{16}\left(e^{2t} - e^{-2t} - 4te^{-2t} \right)$

(ii) $L^{-1}\left\{ \dfrac{s^2}{(s^2+4)^2} \right\} = \dfrac{1}{2}t\cos 2t + \dfrac{1}{4}\sin 2t$

(iii) $L^{-1}\left\{ \dfrac{1}{(s+1)(s^2+1)} \right\} == \dfrac{1}{2}\left(\sin t - \cos t + e^{-t} \right)$

(35) $L^{-1}\left\{ \dfrac{1}{(s+1)(s^2+1)} \right\} = e^{-2t} + e^{-3t}$

(36)   Proof steps required

(37) $Y(t) = (\sin t)/t$

(38) $Y(t) = t$

(39) $Y(t) = e^{2t}$

(40) $Y(t) = t^2 + \dfrac{1}{12} t^4$

(41)

(i) a ≠ b $Y(t) = \dfrac{1}{a-b} \displaystyle\sum_{n=0}^{\infty} \left(a^{n+1} - b^{n+1}\right) \ F(t-nh)U(t-nh)$

(ii) a = b $Y(t) = \displaystyle\sum_{n=0}^{\infty} (n+1) \ a^n \ F(t-nh) \ U(t-nh)$

(42) $Y(t) = t^3 + 6t$

(43) $Y(t) = t^2 / 2$

(44) $Y(t) = t + A t^2$

(45) $Y(t) = C (t - k^2 t^2 / 2)$

(46)   Proof steps required

(47)   Proof steps required

(48)   Proof steps required

(49) $L^{-1}\left\{ \dfrac{2s+1}{s(s^2+1)} \right\} = 1 - \cos t + 2\sin t$

(50)   Proof steps required

(51)   Find $L\left\{ t \ J_1(t) \right\} = \dfrac{1}{(s^2+1)^{\frac{3}{2}}}$

(52) $Y(t) = \cos 2t$

(53)   Solve the difference equations

(i) $Y(t) = t + (t-2)U(t-2) + +(t-4)U(t-4) + ..$

(ii) $Y(t) = \dfrac{t^2}{2!} + \dfrac{(t-1)^3}{3!} U(t-1) + \dfrac{(t-2)^4}{4!} U(t-2) + ..$

187

(54) $$\frac{e^{-cs} - (cs+1)e^{-2cs}}{s^2(1-e^{-2cs})}$$

(55) $$L^{-1}\left\{\frac{1}{s^2(s-1)}\right\} = e^t - (t+1)$$

(56) $$Y(t) = \frac{1}{\sqrt{2}}e^{-t}\sinh\sqrt{2}t$$

(57) $$L^{-1}\left\{\frac{1}{s(s^2-1)}\right\} = 1 + \frac{1}{2}e^t + \frac{1}{2}e^{-t}$$

(58) $$Y(t) = t^2/2$$

(59) $$L\left\{\ F(c,t)\ \right\} = \frac{1 - e^{-cs} + e^{-2cs} - e^{-3cs}}{s(1-e^{-3cs})}$$

(60) $$Y(t) = t^2/2$$

(61) $$Y(t) = t^2 + 4(t-h)^2 U(t-h) + 12(t-2h)^2 U(t-2h) + ..$$

(62) $$L\left\{\ F(c,t)\ \right\} = \frac{2(1 - 2e^{-cs} + 2e^{-2cs} - e^{-3cs})}{s(1-e^{-3cs})}$$

(63) $$Y(t) = c\ (1 - t^2/2)$$

(64) $$L\left\{\ F(c,t)\ \right\} = \frac{c(e^{-\frac{as}{3}} + e^{-\frac{2as}{3}})}{s(1-e^{-as})}$$

(65) $$\int_0^\infty \frac{\cos 2t - \cos 4t}{t}.dt = \log 2$$

(66) $$\int_{0=\infty}^\infty e^{-t}U(t-3)dt = e^{-3}$$

(67) $$L\left\{\ te^{-3t}J_0(t\sqrt{2})\ \right\} = \frac{s+3}{(s^2+6s+11)^{\frac{3}{2}}}$$

(68) $$L^{-1}\left\{\frac{5s-2}{s(s+1)(s11)}\right\} = t - 2 + e^t + e^{-2t}$$

(69) $$Y(t) = \frac{2t^3}{3!} - \frac{2(t-1)^4}{4!}U(t-1) + \frac{2(t-2)^5}{5!}U(t-2) - ..$$

(70) $$L^{-1}\left\{\frac{1}{s^2(s^2+16)}\right\} = \frac{1}{64}\left(4t - \sin 4t\right)$$

CHAPTER 3

ELECTRICAL APPLICATIONS OF THE LAPLACE TRANSFORMATION

3.1. RLC Circuit with charged and uncharged initial conditions

Example 105   Initially Uncharged Circuit

Find the current **I** at time **t** in a circuit consisting of a resistance **R** and inductance **L** in series with a condenser of capacity **C**, when a constant E.M.F is applied at time t = 0. The **initial values** of charge and current being zero.

Solution

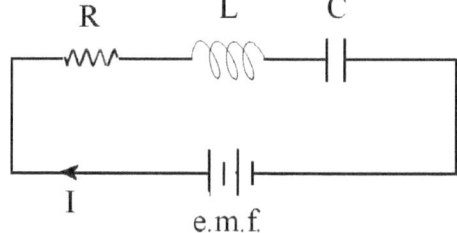

The differential equation satisfied by the current I is

$$I.R + L\frac{dI}{dt} + \frac{1}{C}\int_0^t I(\tau)d\tau = E \qquad (173)$$

With the initial conditions

$$I(0) = 0$$
$$Q(0) = 0$$

Equation (173) is a modified version of **Ohm's equation** that accommodates the active elements L (**inductance**) and C (**capacitance**), in addition to R (the **resistive element**).

The **subsidiary equation** is the Laplace transform of equation (173) and is given by

$$L\left\{ I.R + L\frac{dI}{dt} + \frac{1}{C}\int_0^t I(\tau)d\tau \right\} = L\left\{ E \right\}$$

$$i(s).R + L\left[i(s).s - I(0)\right] + \frac{1}{C}\frac{i(s)}{s} = \frac{E}{s} \qquad (173\text{-}1)$$

Where, the operator L{..} is understood to be the Laplace transform operator, while **L** is the electrical inductance in the above electric circuit.

190

Equation (173-1) is obtained by using the three properties of Laplace Transform of (i) multiplication of the transformed object function by s in case of **derivative of the object function**, (ii) division of the transformed object function by s, in case of **integration of the object function**, and (iii) transform of **constant** by 1/s.

Putting the initial value $I(0) = 0$, equation (173-1) gives

$$i(s).R + L[i(s).s - I(0)] + \frac{1}{C}\frac{i(s)}{s} = \frac{E}{s} \qquad (173\text{-}1)$$

$$i(s) = \frac{E}{s\left(R + s.L + \dfrac{1}{Cs}\right)} \qquad (173\text{-}2)$$

We will manipulate the terms in (173-2) in order to get an elementray function form that could be transformed by known Laplace formulas as follows.

$$i(s) = \frac{E}{sR + s^2.L + \dfrac{1}{C}} = \left(\frac{E}{L}\right)\frac{1}{s^2 + \dfrac{R}{L} + \dfrac{1}{LC}}$$

$$= \left(\frac{E}{L}\right)\frac{1}{(s+m)^2 + n^2} \qquad (173\text{-}3)$$

Where the two constants m and n are obtained by equating the coefficients of equal powers of s in the polynomials $s^2 + \dfrac{R}{L}s + \dfrac{1}{LC}$ and $(s^2 + 2\,ms + m^2 + n^2)$. Therefore,

$$m = R/2L \qquad (173\text{-}3.1)$$
$$m^2 + n^2 = 1/LC$$

$$n^2 = \frac{1}{CL} - \left(\frac{R}{2L}\right)^2 \qquad (173\text{-}3.2)$$

From (173-3.2) we notice that the inverse Laplace transform of (173-3) depends on the sign of n^2 such that we have three distinct behaviors of the electric current I(t) for the three values of n^2 = +v, 0, and –ve.

(i) Case n^2 is positive

$$I(t) = L^{-1}\left[\left(\frac{E}{L}\right)\frac{1}{(s+m)^2 + n^2}\right] = \frac{E}{nL}L^{-1}\left[\frac{n}{(s+m)^2 + n^2}\right]$$

$$= \frac{E}{nL}e^{-mt}\sin nt \qquad (173\text{-}4)$$

Substituting by the values of m and n from (173-3.1) and (173-3.2), we get

$$I(t) = \frac{E}{L\sqrt{\frac{1}{CL} - \left(\frac{R}{2L}\right)^2}} e^{-\frac{Rt}{2L}} \sin\left(\sqrt{\frac{1}{CL} - \left(\frac{R}{2L}\right)^2}\; t\right) \qquad (173\text{-}4)$$

(ii) Case n^2 is zero

From (173-3.2), we get

$$0 = \frac{1}{CL} - \left(\frac{R}{2L}\right)^2$$

Or $\qquad R = 2\sqrt{\dfrac{L}{C}}$ $\qquad\qquad\qquad (173\text{-}5.1)$

Equation (173-4) gives

$$I(t)_{n=0} = \frac{E}{L} e^{-mt} \lim_{n\to 0} \frac{\sin nt}{n}$$

$$I(t)_{n=0} = \frac{E}{L} e^{-mt}\; t \lim_{n\to 0} \frac{\sin nt}{nt} = \frac{E}{L} e^{-mt}\; t$$

$$I(t)_{n=0} = \frac{E}{L} e^{-\frac{R}{2L}t}\; t \qquad\qquad\qquad (173\text{-}5.2)$$

(iii) Case n^2 is negative

Let us substitute in equation (173-3) by $n^2 = -\mu^2$, where μ is positive. Therefore

$$I(t) = L^{-1}\left\{\left(\frac{E}{L}\right)\frac{1}{(s+m)^2 - \mu^2}\right\} = \frac{E}{\mu L} L^{-1}\left\{\frac{\mu}{(s+m)^2 - \mu^2}\right\}$$

$$= \frac{E}{\mu L} e^{-mt} \sinh \mu t$$

$$= \frac{E}{L\sqrt{-\frac{1}{CL} + \left(\frac{R}{2L}\right)^2}} e^{-\frac{Rt}{2L}} \sinh\left(\sqrt{-\frac{1}{CL} + \left(\frac{R}{2L}\right)^2}\; t\right) \qquad (173\text{-}6)$$

The three cases of temporal variation of the electric current in equations (173-4), (173-5), and (173-6) are shown in the following graph.

Case	R	C	L	nn	m	n	mu
1	1	1	4	0			
2	1	1	1	0.75	0.5	0.866025	
3	100	0.001	10	75	5	8.660254	
4	1000	0.001	10	-2400	50		48.98979
5	1000	0.1	2	-62495	250		249.99

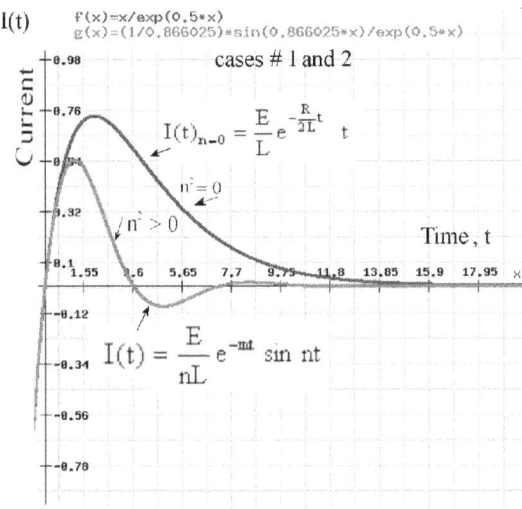

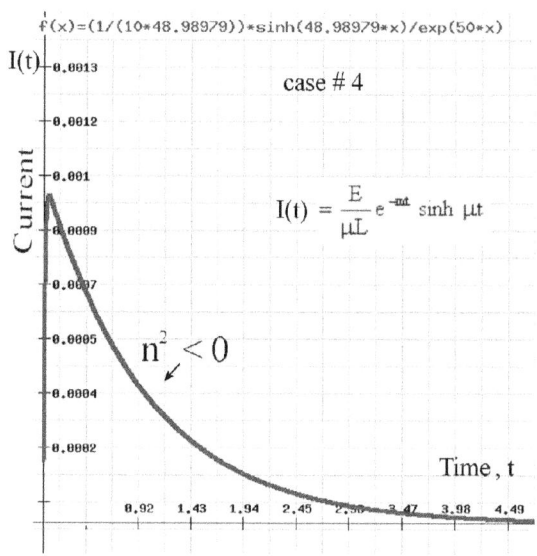

f(x)=(1/(10*48.98979))*sinh(48.98979*x)/exp(50*x)

I(t)

case # 4

$$I(t) = \frac{E}{\mu L} e^{-mt} \sinh \mu t$$

$n^2 < 0$

Current

Time, t

0.92   1.43   1.94   2.45   2.96   3.47   3.98   4.49

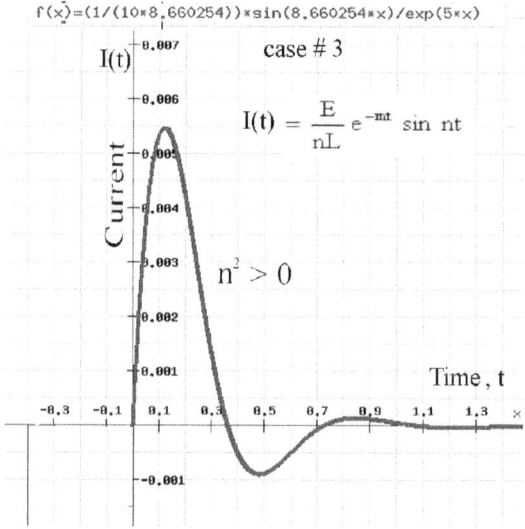

f(x)=(1/(10*8.660254))*sin(8.660254*x)/exp(5*x)

I(t)   case # 3

$$I(t) = \frac{E}{nL} e^{-mt} \sin nt$$

$n^2 > 0$

Current

Time, t

-0.3   -0.1   0.1   0.3   0.5   0.7   0.9   1.1   1.3   x

194

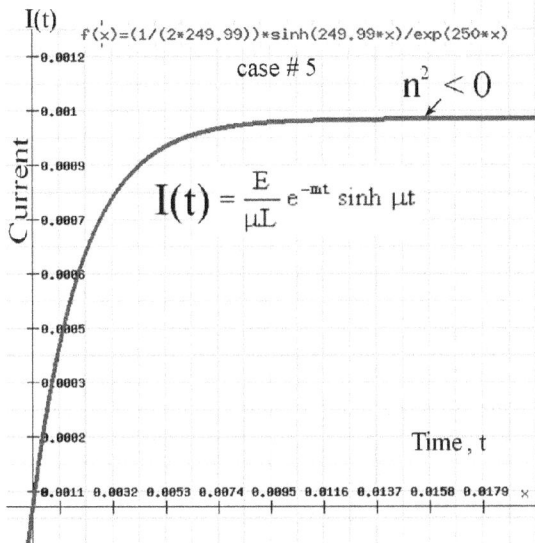

$$f(x)=(1/(2*249.99))*\sinh(249.99*x)/\exp(250*x)$$

case # 5

$$n^2 < 0$$

$$I(t) = \frac{E}{\mu L} e^{-mt} \sinh \mu t$$

Time, t

0.0011 0.0032 0.0053 0.0074 0.0095 0.0116 0.0137 0.0158 0.0179 x

Example 106 Initially Charged Circuit

Find the current in a series circuit of resistance R, inductance L and capacity C in which an E.M.F. V is applied at t = 0, **the initial values** of the current I and charge Q on the condenser being respectively I_o and Q_o.

Solution

Equation (173), can be written as follows

$$I.R + L\frac{dI}{dt} + \frac{Q}{C} = V \tag{174}$$

Where, the current I(t) and the charge Q(t) are connected by

$$I(t) = dQ(t) / dt \tag{174-1}$$

The **subsidiary equations** of (174) and (174-1) are the Laplace transforms obtained as before, as follows:

$$i(s).R + L[i(s).s - I(0)] + \frac{q(s)}{C} = \frac{V}{s} \tag{174-2}$$

$$i(s) = sq(s) - Q(0) \tag{174-3}$$

195

In contrast to equations (173-1) for initially uncharged circuit, here we have inial **charge** Q(0) and **current** I(0).

Eliminating q(s) we get

$$i(s).R + L[i(s).s - I(0)] + \frac{1}{Cs}[i(s) + Q(0)] = \frac{V}{s} \tag{174-4}$$

$$\left(L.s + R + \frac{1}{Cs}\right) i(s) = \frac{V}{s} + L.I_0 + \frac{Q_0}{Cs} \tag{174-5}$$

$$i(s) = \frac{\dfrac{V}{s} + L.I_0 + \dfrac{Q_0}{Cs}}{L.s + R + \dfrac{1}{Cs}}$$

$$= \frac{\dfrac{V}{L} + \dfrac{Q_0}{CL} + s.I_0}{.s^2 + s.\dfrac{R}{L} + \dfrac{1}{CL}} = \frac{\dfrac{V}{L} + \dfrac{Q_0}{CL}}{.s^2 + s.\dfrac{R}{L} + \dfrac{1}{CL}} + \frac{s.I_0}{.s^2 + s.\dfrac{R}{L} + \dfrac{1}{CL}}$$

$$= \frac{w}{.(s+m)^2 + n^2} + \frac{s.I_0}{.(s+m)^2 + n^2} \tag{174-6}$$

Where, the constants m, n, and w are defined as:

$$w = \frac{V}{L} + \frac{Q_0}{CL} \tag{174-6.1}$$

$$m = R / 2L \tag{174-6.2}$$

$$n^2 = \frac{1}{CL} - \left(\frac{R}{2L}\right)^2 \tag{174-6.3}$$

The inverse Laplace Transform of (174-6) is obtained from the elementary sine and cosine functions as follows.

sin kt	L{sin kt}	$\dfrac{k}{k^2 + s^2}$	
cos kt	L{cos kt}	$\dfrac{s}{k^2 + s^2}$	, s > 0

Therefore, equation (174-6) gives

$$I(t) = L^{-1}\left(\frac{w}{.(s+m)^2 + n^2} + \frac{s.I_0}{.(s+m)^2 + n^2}\right)$$

$$= L^{-1}\left(\frac{w}{n}\frac{n}{.(s+m)^2+n^2}+I_o\frac{s}{.(s+m)^2+n^2}\right)$$

$$= \frac{w}{n}e^{-mt}\sin nt + I_o e^{-mt}\cos mt \qquad\qquad (174\text{-}7)$$

********** Auxiliary Proof sin ix = sin x and cos ix = cosh x **************

From the series expansions of sine, cosine, sinh, and cosh, we can prove that

$$\sin ix = \sinh x \qquad\qquad (174\text{-}7.1)$$
$$\cos ix = \cosh x \qquad\qquad (174\text{-}7.2)$$

From (108-4.4) through (108-4.7), we have the following series expansions:

$$\cos x = \sum_{n=0}^{\infty}\frac{(-1)^n x^{2n}}{(2n)!} \qquad \text{all values of x} \qquad (174\text{-}7.3)$$

$$\sin x = \sum_{n=0}^{\infty}\frac{(-1)^n x^{2n+1}}{(2n+1)!} \qquad \text{all values of x} \qquad (174\text{-}7.4)$$

$$\cosh x = \sum_{n=0}^{\infty}\frac{x^{2n}}{(2n)!} \qquad \text{all values of x} \qquad (174\text{-}7.5)$$

$$\sinh x = \sum_{n=0}^{\infty}\frac{x^{2n+1}}{(2n+1)!} \qquad \text{all values of x} \qquad (174\text{-}7.6)$$

Substituting by ix in (174-7.3), we get

$$\cos ix = \sum_{n=0}^{\infty}\frac{(-1)^n (ix)^{2n}}{(2n)!} = \sum_{n=0}^{\infty}\frac{(-1)^n (i)^{2n}(x)^{2n}}{(2n)!} = \sum_{n=0}^{\infty}\frac{(-1)^n (-1)^n (x)^{2n}}{(2n)!}$$

$$= \sum_{n=0}^{\infty}\frac{(x)^{2n}}{(2n)!} = \cosh x \qquad\qquad (174\text{-}7.7)$$

Similarly, by substituting ix in (174-4.4) we get **sin ix = sinh x.**

********** End Auxiliary Proof sin ix = sin x and cos ix = cosh x **************

From equation (174-7), we have three states depending on the sign of n.

(i) Case n^2 is positive

$$I(t) = e^{-\frac{R}{2L}t}\left(\frac{\frac{V}{L}+\frac{Q_o}{CL}}{\sqrt{\frac{1}{CL}-\left(\frac{R}{2L}\right)^2}}\sin\left(\sqrt{\frac{1}{CL}-\left(\frac{R}{2L}\right)^2}\;t\right)+I_o\cos\left(\sqrt{\frac{1}{CL}-\left(\frac{R}{2L}\right)^2}\;t\right)\right) \qquad (174\text{-}8)$$

(ii) Case n^2 is zero

$$I(t) = e^{-\frac{R}{2L}t}\left(\left(\frac{V}{L}+\frac{Q_o}{CL}\right)\;t+I_o\right) \qquad (174\text{-}9)$$

(iii) Case n^2 is negative

$$I(t) = e^{-\frac{R}{2L}t}\left(\frac{\frac{V}{L}+\frac{Q_o}{CL}}{\sqrt{\left(\frac{R}{2L}\right)^2-\frac{1}{CL}}}\sinh\left(\sqrt{\left(\frac{R}{2L}\right)^2-\frac{1}{CL}}\;t\right)+I_o\cosh\left(\sqrt{\left(\frac{R}{2L}\right)^2-\frac{1}{CL}}\;t\right)\right) \qquad (174\text{-}10)$$

Substituting by $Io = Qo = 0$ in equations (174-8), (174-9), and (174-10), we get, respectively, (173-4), (173-5.2), and (173-6).

Example 107   Abruptly Shorted Battery

A battery of E M.F. E is connected at time $t = 0$ to a series circuit of resistance R., inductance L and capacity C. The **initial values** of the current and charge are zero. If the **battery is short-circuited** at $t = T$. Find the current I at any instant t.

Solution

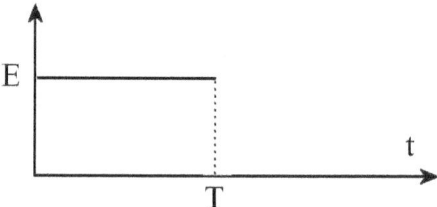

The modified **Ohm's equation** that accommodates the active elements L (**inductance**) and C (**capacitance**), in addition to R (the **resistive element**) is

198

$$I.R + L\frac{dI}{dt} + \frac{1}{C}\int_0^t I(\tau)d\tau = V(t) \tag{175}$$

With the initial conditions

$$
\begin{aligned}
V(t) &= E & & 0 < t < T \\
&= 0 & & t > T
\end{aligned}
\tag{175-1}
$$

The **subsidiary equation** is the Laplace transform of equations (175) and (175-1) and is given by

$$L\left\{ I.R + L\frac{dI}{dt} + \frac{1}{C}\int_0^t I(\tau)d\tau \right\} = L\left\{ V \right\} = v(s)$$

$$i(s).R + L[i(s).s - I(0)] + \frac{1}{C}\frac{i(s)}{s} = v(s) \tag{175-2}$$

And $\qquad v(s) = \frac{E}{s}\left(1 - e^{-Ts}\right) \tag{175-3}$

From (175-2) and (175-3), the Laplace transform of electric current is

$$
\begin{aligned}
i(s) &= \frac{1}{R + L[s - I(0)] + \dfrac{1}{Cs}}\left(\frac{E}{s}\right)\left(1 - e^{-Ts}\right) \\[2mm]
&= \left(\frac{E}{L}\right)\frac{1 - e^{-Ts}}{s^2 + s\dfrac{R}{L} + \dfrac{1}{CL}} \\[2mm]
&= \left(\frac{E}{L}\right)\left(\frac{1}{(s+m)^2 + n^2} - \frac{e^{-Ts}}{(s+m)^2 + n^2}\right) \\[2mm]
&= \left(\frac{E}{nL}\right)\left(\frac{n}{(s+m)^2 + n^2} - \frac{ne^{-Ts}}{(s+m)^2 + n^2}\right)
\end{aligned}
\tag{175-4}
$$

The inverse Laplace Transform of (175-4) yields the temporal profile of electric current

$$
\begin{aligned}
I(t) &= \left(\frac{E}{nL}\right)L^{-1}\left\{\frac{n}{(s+m)^2 + n^2} - \frac{ne^{-Ts}}{(s+m)^2 + n^2}\right\} \\[2mm]
&= \left(\frac{E}{nL}\right)e^{-mt}\left(\sin nt - U(t - T).\sin n(t - T)\right)
\end{aligned}
\tag{175-5}
$$

We have used the **unit-step function**, equation (57-3), which implies

$$L^{-1}\left\{e^{-st} f(s)\right\} = U(t - a)F(t - a), \quad s > 0 \tag{175-5.1}$$

At $0 < t < T$, $U(t-T) = 0$

Therefore, equations (175-5) the electric current is

$$I(t) = \left(\frac{E}{nL}\right) e^{-mt} \sin nt$$

$$= \left(\frac{E}{L\sqrt{\frac{1}{LC} - \left(\frac{R}{2L}\right)^2}}\right) e^{-\frac{R}{2L}t} \sin\left(\sqrt{\frac{1}{LC} - \left(\frac{R}{2L}\right)^2}\ t\right) \tag{175-6}$$

At $t > T$, $U(t-T) = 1$

Therefore, equations (175-5) the electric current is

$$I(t) = \left(\frac{E}{nL}\right)\left(e^{-mt} \sin nt - e^{-m(t-T)} \sin n(t-T)\right)$$

$$= \left(\frac{Ee^{-\frac{R}{2L}t}}{L\sqrt{\frac{1}{LC} - \left(\frac{R}{2L}\right)^2}}\right)\left(\sin\left(\sqrt{\frac{1}{LC} - \left(\frac{R}{2L}\right)^2}\ t\right) - e^{\frac{R}{2L}T} \sin\left(\sqrt{\frac{1}{LC} - \left(\frac{R}{2L}\right)^2}(t-T)\right)\right) \tag{175-7}$$

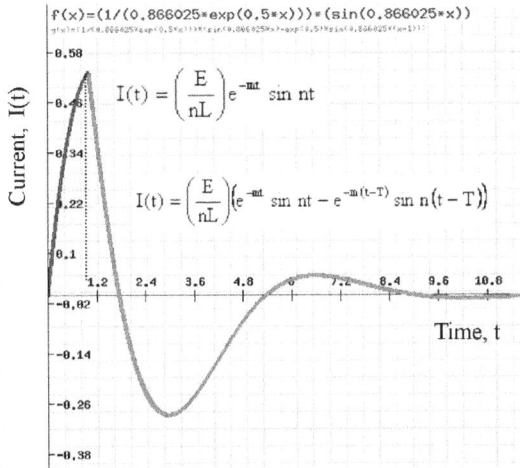

f(x)=(1/(0.866025*exp(0.5*x)))*(sin(0.866025*x))

$$I(t) = \left(\frac{E}{nL}\right)e^{-mt}\,\sin\,nt$$

$$I(t) = \left(\frac{E}{nL}\right)\left(e^{-mt}\,\sin\,nt - e^{-m(t-T)}\,\sin\,n(t-T)\right)$$

Example 108   Resonance CL Circuit

A periodic E. M. F. of resonance frequency **E sin(nt)** is applied at time t = 0 to a series circuit consisting of inductance **L** and capacity **C,** where $n^2 = 1 / CL$.

Find the current in the circuit at time t, assuming initial zero current and charge.

Solution

The modified **Ohm's equation** that accommodates the active elements L (**inductance**) and C (**capacitance**), without R (the **resistive element**) is

$$L\frac{dI}{dt} + \frac{1}{C}\int_0^t I(\tau)d\tau = E\sin\,nt \qquad (176)$$

With the conditions I(0) = 0, Q (0) = 0

$$i(s).s.L + \frac{1}{C}\frac{i(s)}{s} = \frac{nE}{s^2 + n^2} \qquad (176\text{-}1)$$

$$i(s) = \frac{nE}{\left(s.L + \dfrac{1}{Cs}\right)\left(s^2 + n^2\right)}$$

$$= \frac{nE}{\dfrac{L}{s}\left(s^2 + \dfrac{1}{CL}\right)\left(s^2 + n^2\right)} = \frac{s.nE}{L\left(s^2 + n^2\right)\left(s^2 + n^2\right)} = \frac{nE}{L}\frac{s}{\left(s^2 + n^2\right)^2} \qquad (176\text{-}2)$$

The inverse Laplace Transform of (176-2)

$$I(t) = L^{-1}\left\{\frac{nE}{L}\frac{s}{\left(s^2 + n^2\right)^2}\right\} = \frac{nE}{L}\cdot\frac{1}{2n}t\sin nt$$

$$= \frac{E}{2L}t\ \sin nt \qquad (176\text{-}3)$$

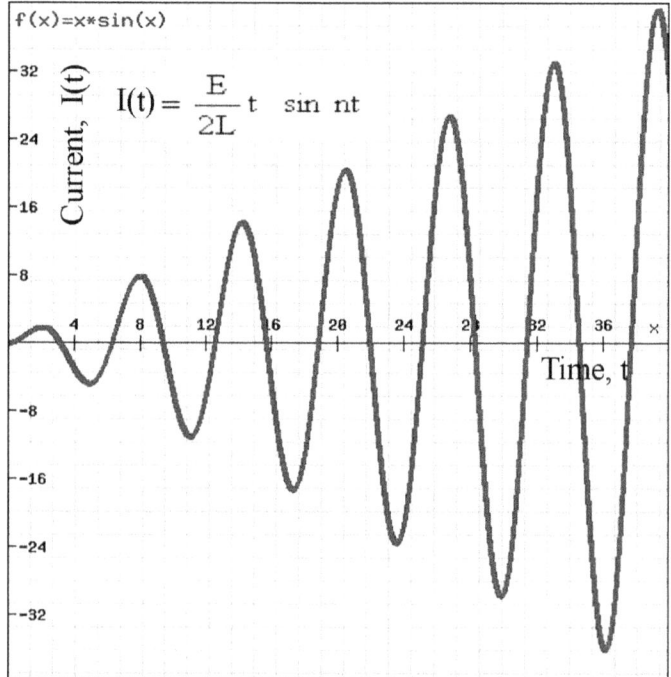

f(x)=x*sin(x)

$$I(t) = \frac{E}{2L}t\ \sin\ nt$$

Current, I(t)

Time, t

Example 109   **Resonance RL Circuit**

Find the current at time t when a periodic E.M.F. **E sin** *wt* is applied at time t = 0 to an inductive resistance L, R the initial current being zero.

Solution

$$L\frac{dI}{dt} + RI = E \sin \omega t \tag{177}$$

With the conditions $I(0) = 0$, $Q(0) = 0$.

$$i(s).s.L + i(s)R = \frac{\omega E}{s^2 + \omega^2} \tag{177-1}$$

$$i(s) = \frac{\omega E}{(s.L + R)(s^2 + \omega^2)} = \frac{\omega E}{L\left(s + \dfrac{R}{L}\right)(s^2 + \omega^2)} \tag{177-2}$$

Substitute by $\mu = R / L$, we get

$$i(s) = \frac{\omega E}{L(s + \mu)(s^2 + \omega^2)} \tag{177-3}$$

Equation (177-3) is written in its partial fraction form as follows

$$i(s) = \frac{\omega E}{L(\mu^2 + \omega^2)}\left(\frac{1}{s + \mu} - \frac{s - \mu}{s^2 + \omega^2}\right) \tag{177-4}$$

The inverse Laplace transform of (177-4) is

$$\begin{aligned}
I(t) &= \frac{\omega E}{L(\mu^2 + \omega^2)}L^{-1}\left\{\frac{1}{s + \mu} - \frac{s - \mu}{s^2 + \omega^2}\right\} \\
&= \frac{\omega E}{L(\mu^2 + \omega^2)}\left(e^{-\mu t} - \cos \omega t + \frac{\mu}{\omega}\sin \omega t\right)
\end{aligned} \tag{177-5}$$

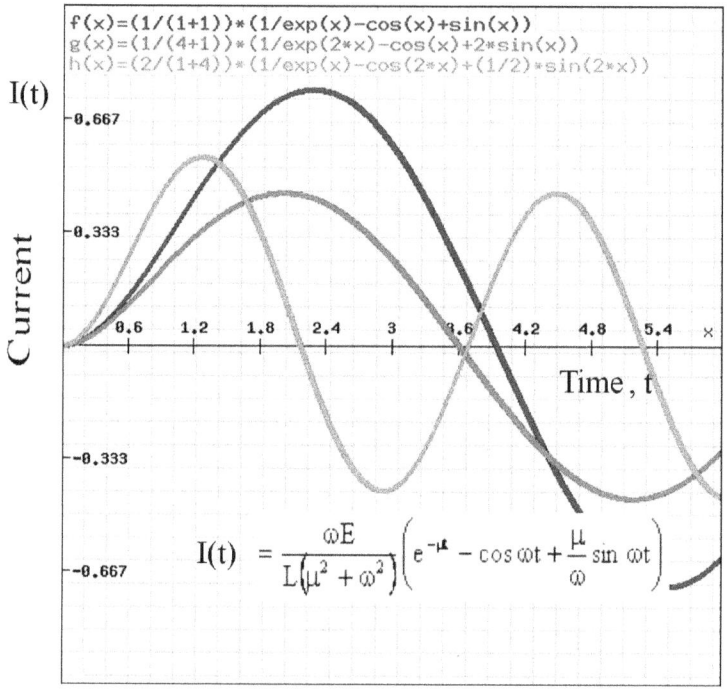

```
f(x)=(1/(1+1))*(1/exp(x)-cos(x)+sin(x))
g(x)=(1/(4+1))*(1/exp(2*x)-cos(x)+2*sin(x))
h(x)=(2/(1+4))*(1/exp(x)-cos(2*x)+(1/2)*sin(2*x))
```

$$I(t) = \frac{\omega E}{L(\mu^2 + \omega^2)}\left(e^{-\mu t} - \cos \omega t + \frac{\mu}{\omega}\sin \omega t\right)$$

Example 110   **RCL Circuit Transition of Potentials**

Application 6.

An E.M.F. E_1 for $0 < t < T$ and E_2 for $t > T$, where E_1 and E_2 are constants is applied to a series circuit L, R, C. Find the current at any time t assuming zero initial current and charge.

Solution

The modified **Ohm's equation** for the RCL circuit is

$$I.R + L\frac{dI}{dt} + \frac{1}{C}\int_0^t I(\tau)d\tau = V(t) \tag{178}$$

With the initial conditions

$$\begin{aligned} V(t) &= E_1 & 0 < t < T \\ &= E_2 & t > T \end{aligned} \tag{178-1}$$

The **subsidiary equation** is the Laplace transform of equations (178-1) is given by

204

$$L\left\{\ V(t)\ \right\} = \frac{E_1}{s} - \frac{E_1 - E_s}{s}e^{-Ts}$$ (178-2)

We have assumed that the voltage drop E_1 - E_2 applied at time $t = T$ is a step function shifted T from the $t = 0$. Therefore, (178) becomes

$$i(s).R + L.i(s).s + \frac{1}{C}\frac{i(s)}{s} = \frac{E_1}{s} - \frac{E_1 - E_s}{s}e^{-Ts}$$ (178-3)

$$= \left(\frac{E_1}{L}\right)\frac{1}{s^2 + s\frac{R}{L} + \frac{1}{CL}} - \left(\frac{E_1 - E_s}{L}\right)\frac{e^{-Ts}}{s^2 + s\frac{R}{L} + \frac{1}{CL}}$$

$$= \left(\frac{E_1}{nL}\right)\frac{n}{(s+m)^2 + n^2} - \left(\frac{E_1 - E_2}{nL}\right)\left(-\frac{ne^{-Ts}}{(s+m)^2 + n^2}\right)$$ (178-4)

The inverse Laplace Transform of (178-4) yields the temporal profile of electric current

$$I(t) = L^{-1}\left\{\left(\frac{E_1}{nL}\right)\frac{n}{(s+m)^2 + n^2} - \left(\frac{E_1 - E_2}{nL}\right)\frac{ne^{-Ts}}{(s+m)^2 + n^2}\right\}$$

$$= e^{-mt}\left[\left(\frac{E}{nL}\right)\sin nt - \left(\frac{E_1 - E_2}{nL}\right)e^{mT}U(t-T).\sin n(t-T)\right]$$ (178-5)

Again, we have used the **unit-step function**, equation (57-3), which implies

$$L^{-1}\left\{e^{-st}\ f(s)\right\} = U(t-a)F(t-a),\quad s > 0$$ (178-5.1)

At $0 < t < T$, $U(t-T) = 0$

Equation (178-5.1) gives

$$I(t) = e^{-mt}\left(\frac{E}{nL}\right)\sin nt$$ (178-6)

At $t > T$, $U(t-T) = 1$

Equation (178-5.1) gives

$$I(t) = e^{-mt}\left[\left(\frac{E}{nL}\right)\sin nt - \left(\frac{E_1 - E_2}{nL}\right)e^{mT}\sin n(t-T)\right]$$ (178-7)

3.2. Mutual induction circuits

Example 111   **Primary and secondary induction**

The two circuits shown in the diagram are coupled by **mutual inductance** M. A constant E.M.F. E is applied to the **primary circuit** at time t = 0 with zero initial conditions, the primary resistance being neglected. Find **secondary current** at any time t.

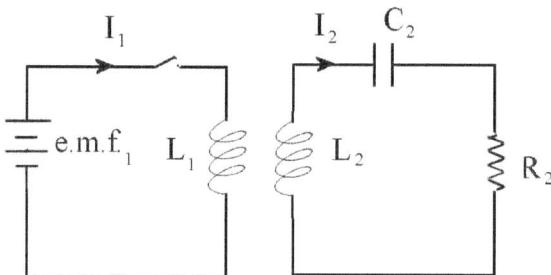

For the primary circuit, Ohm's law gives

$$I.0 + L_1\frac{dI_1}{dt} + M\frac{dI_2}{dt} + \frac{0}{C}\int_0^t I(\tau)d\tau = E \tag{179}$$

With the initial conditions

$$I_1(0) = 0$$
$$Q_1(0) = 0$$

For the secondary, Ohm's law gives

$$I_2.R_2 + L_2\frac{dI_2}{dt} + M\frac{dI_1}{dt} + \frac{1}{C}\int_0^t I_2(\tau)d\tau = 0 \tag{179-1}$$

With the initial conditions

$$I_2(0) = 0$$
$$Q_2(0) = 0$$

The subsidiary equations are the Laplace transforms of (179) and (179-1) and are given as follows

$$L_1i_1(s).s + M.i_2(s).s = \frac{E}{s} \tag{179-2}$$

$$R_2 i_2(s) + L_2 . i_2(s).s + M.i_1(s).s + \frac{1}{C_2} \frac{i_2(s)}{s} = 0 \qquad (179\text{-}3)$$

Rearranging equations (179-2) and (179-3), we get

$$i_1(s) = \frac{\dfrac{E}{s} - i_2(s).s.M}{s..L_1} = \frac{E}{s^2 L_1} - i_2(s).\frac{M}{L_1} \qquad (179\text{-}4)$$

$$i_2(s)\left(R_2 + L_2.s + \frac{1}{sC_2} \right) + M.i_1(s).s = 0 \qquad (179\text{-}5)$$

Substituting by (179-4) in (179-5), we get

$$i_2(s)\left(R_2 + L_2.s + \frac{1}{sC_2} \right) + M.s\left(\frac{E}{s^2 L_1} - i_2(s).\frac{M}{L_1} \right) = 0 \qquad (179\text{-}6)$$

Thus, from (179-6) the Laplace transform of the secondary current is given by

$$i_2(s)\left(R_2 + L_2.s + \frac{1}{sC_2} - \frac{M^2}{L_1}.s \right) = -\frac{M.E}{sL_1}$$

$$i_2(s) = -\frac{M.E}{sL_1\left(R_2 + L_2.s + \dfrac{1}{sC_2} - \dfrac{M^2}{L_1}.s \right)}$$

$$= -\frac{M.E}{L_1}.\frac{1}{\left(L_2. - \dfrac{M^2}{L_1} \right)s^2 + sR_2 + \dfrac{1}{C_2}}$$

$$= -\left(\frac{M.E}{L_1 L_2 - M^2} \right).\frac{1}{s^2 + s\dfrac{L_1 R_2}{L_1 L_2 - M^2} + \dfrac{L_1}{C_2\left(L_1 L_2 - M^2 \right)}}$$

$$= -\left(\frac{M.E}{L_1 L_2 - M^2} \right).\frac{1}{(s+m)^2 + n^2} \qquad (179\text{-}7)$$

And, substituting from (179-7) in (179-4), we get

$$i_1(s) = \frac{E}{s^2 L_1} + \frac{M}{L_1}\left(\frac{M.E}{L_1 L_2 - M^2} \right).\frac{1}{(s+m)^2 + n^2} \qquad (179\text{-}8)$$

Where, the constants m and n substitute the constants of the circuits according to the definitions:

$$m = \frac{1}{2}\left(\frac{L_1 R_2}{L_1 L_2 - M^2}\right)$$ (179-9)

$$n^2 = \frac{L_1}{C_2(L_1 L_2 - M^2)} - m^2 = \frac{L_1}{C_2(L_1 L_2 - M^2)} - \frac{1}{4}\left(\frac{L_1 R_2}{L_1 L_2 - M^2}\right)^2$$ (179-10)

The termporal profile of the primary and secondary currents are the inverse Laplace transforms of (179-7) and (179-8), as follows.

$$I_2(t) = -\left(\frac{M.E}{L_1 L_2 - M^2}\right).L^{-1}\left(\frac{1}{(s+m)^2 + n^2}\right)$$

$$= -\frac{1}{n}\left(\frac{M.E}{L_1 L_2 - M^2}\right).e^{-mt} \sin nt$$ (179-11)

$$I_1(t) = L^{-1}\left\{\frac{E}{s^2 L_1} + \frac{M}{L_1}.\left(\frac{M.E}{L_1 L_2 - M^2}\right).\frac{1}{(s+m)^2 + n^2}\right\}$$

$$= \frac{E}{L_1}t + \frac{M}{nL_1}.\left(\frac{M.E}{L_1 L_2 - M^2}\right).e^{-mt} \sin nt$$ (179-12)

L_1 L_2 R_2 C_2 E	M	n	m	$A =$ $-\dfrac{1}{n}\left(\dfrac{M.E}{L_1 L_2 - M^2}\right).$	$B =$ $\dfrac{E}{L_1}$	$C=$ $\dfrac{M}{nL_1}\left(\dfrac{M.E}{L_1 L_2 - M^2}\right)$
1	0.8	0.921285	1.388889	-2.41209	1	1.929673
1	0.2	0.877724	0.520833	-0.23736	1	0.047471

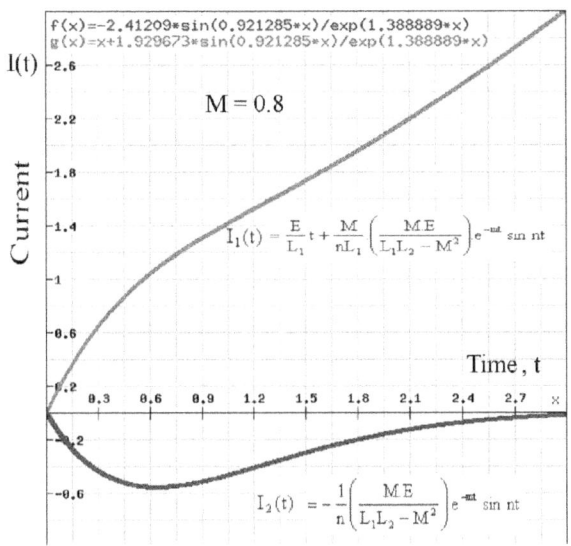

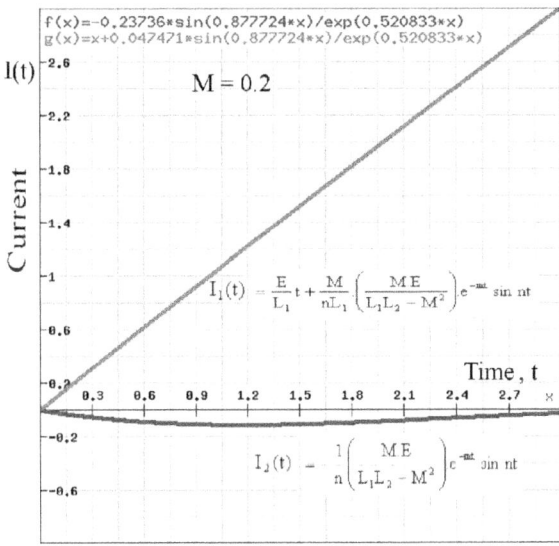

Example 112   **RCL Networks**

In the following network find the total current I at any instant t, the initial currents and charges are zero.

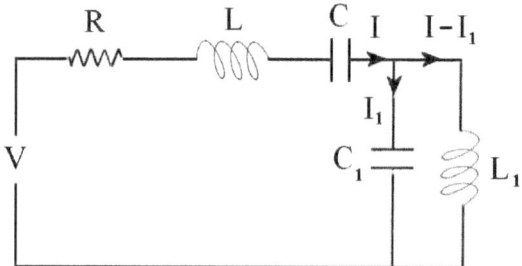

Solution

Applying Kirchoff's voltage law to the circuit on the left which states that the impressed voltage equate the sum of the voltage drops across the elements in that circuit we get

$$I.R + L\frac{dI}{dt} + \frac{1}{C}\int_0^t I(\tau)d\tau + \frac{1}{C_1}\int_0^t I_1(\tau)d\tau = V \tag{180}$$

Again, applying the Kirchoff's voltage law to the circuit on the right we get

$$I_1.0 + L\frac{d(I - I_1)}{dt} = \frac{1}{C_1}\int_0^t I_1(\tau)d\tau \tag{180-1}$$

The subsidiary equations are the Laplace transforms of (180) and (180-1) and are obtained as follows

$$i(s).R + L[i(s).s - I(0)] + \frac{i(s)}{sC} + \frac{i_1(s)}{C_1 s} = v(s) \tag{180-2}$$

$$L\left[i(s).s - i_1(s).s\right] - \frac{i_1(s)}{C_1 s} = 0 \tag{180-3}$$

Arranging the terms in (180-2) and (180-3), we get

(i) For the primary circuit, (180-2) gives

$$i(s)\left(R + L.s + \frac{1}{sC}\right) + \frac{1}{C_1 s}i_1(s) = v(s)$$

$$i(s)\left(R + L.s + \frac{1}{sC}\right) + \frac{1}{C_1 s}i_1(s) = v(s) \tag{180-4}$$

(2) For the secondary circuit, (180-3) gives

$$i_1(s) = \frac{L_1 \; i(s).s}{L_1.s + \dfrac{1}{C_1 s}} \tag{180-5}$$

Substituting from (180-5) into (180-4) we get

$$i(s)\left(R + L.s + \frac{1}{sC}\right) + \left(\frac{1}{C_1 s}\right)\frac{L_1 \; i(s).s}{L_1.s + \dfrac{1}{C_1 s}} = v(s)$$

$$i(s)\left(R + L.s + \frac{1}{sC}\right) + \frac{L_1 \; i(s)s}{C_1 L_1.s^2 + 1} = v(s)$$

$$i(s)\left(R + L.s + \frac{1}{sC} + \frac{L_1 \; s}{C_1 L_1.s^2 + 1}\right) = v(s)$$

Therefore, the Laplace transform of the **primary current** is

$$i(s) = \frac{v(s)}{\left(R + L.s + \dfrac{1}{sC} + \dfrac{L_1 \; s}{C_1 L_1.s^2 + 1}\right)} \tag{180-6}$$

The secondary current is obtained by substituting from (180-6) into (180-5) to get

$$i_1(s) = \frac{L_1 \; s}{L_1.s + \dfrac{1}{C_1 s}} \cdot \frac{v(s)}{\left(R + L.s + \dfrac{1}{sC} + \dfrac{L_1 \; s}{C_1 L_1.s^2 + 1}\right)} \tag{180-7}$$

Laplace inverting of (180-6) and (180-7) we get the required currents in the two circuits.

***** Proof by Partial Fractions ************

Let us consider the case of constant voltage. Thus

$$v(s) = V/s \tag{180-7.1}$$

Equation (180-6) becomes

$$i(s) = \cfrac{V}{s \left[R + L.s + \cfrac{1}{sC} + \cfrac{L_1 \, s}{C_1 L_1 . s^2 + 1} \right]} = \cfrac{V}{sR + L.s^2 + \cfrac{1}{C} + \cfrac{L_1 s^2}{C_1 L_1 . s^2 + 1}}$$

$$= \frac{VC(C_1 L_1 . s^2 + 1)}{C(sR + L.s^2) + (C_1 L_1 . s^2 + 1) + (C_1 L_1 . s^2 + 1) + CL_1 s^2}$$

$$= \frac{VC(C_1 L_1 . s^2 + 1)}{(C(sR + L.s^2) + 1)(C_1 L_1 . s^2 + 1) + CL_1 s^2} \qquad (180\text{-}7.2)$$

We will arrange the terms in (180-7.2) in the form of 4th power polynomial as follows

$$i(s) = \frac{VC(C_1 L_1 . s^2 + 1)}{s^4 (CLC_1 L_1) + s^3 (CRC_1 L_1) + s^2 (CL + C_1 L_1 + CL_1) + sCR + C} \qquad (180\text{-}7.3)$$

$$= \frac{VC(k.s^2 + 1)}{m s^4 + n s^3 + p s^2 + w.s + C} \qquad (180\text{-}7.4)$$

$$= \frac{VC(k.s^2 + 1)}{(s^2 + A^2)(s + B)(s + D)} \qquad (180\text{-}7.5)$$

Where, the constants k, m, n, p, and w are given by the coefficients of s^4, s^3, s^2, s, and constants in equation (180-7.3)

$k = C_1 L_1$	(180-7.5.1)
$m = CLC_1 L_1$	(180-7.5.2)
$n = CRC_1 L_1$	(180-7.5.3)
$p = CL + C_1 L_1 + CL_1$	(180-7.5.4)
$w = CR$	(180-7.5.5)

Also, the constants A, B, and D in equation (180-7.5) are obtained by equating the coefficients of equal powers of s as follows.

(i) $m = 1$ Coefficients of s^4
(ii) $n = D + B$ Coefficients of s^3

$B = n - D$ (180-7.6)

(iii) $p = BD + A^2$ Coefficients of s^2

From (180-7.6) and (iii) we get

$p = (n - D) D + A^2$

$A^2 = p - (n - D) D$ (180-7.7)

(vi) $w = A^2 D + A^2 B$ Coefficients of s

$A^2 = w / (D + B) = w/n$ (180-7.8)

(v) $C = A^2 B D$ Constants in (180-7.4)

From (180-7.8) and (v) we get

$BD = C/A^2 = C n / w$ (180-7.9)

From (iii), substitute by A^2 and BD to get

$p = (n - D) D + A^2$

$D^2 - p n D - w / n = 0$ (180-7.10)

The quadratic equation in (180-7.10) is solved for D to give

$$D = \frac{1}{2}\left(pn \pm \sqrt{(pn)^2 + 4w/n} \right)$$ (180-7.11)

From (180-7.6) and (180-7.11), we get

$$B = n - \frac{1}{2}\left(pn \pm \sqrt{(pn)^2 + 4w/n} \right)$$ (180-7.12)

Thus, with the constants A, B, and D, determined by (180-7.8), (180-7.11), and (180-7.12), we could write (180-7.4) in the form of partial fractions as follows.

$$i(s) = \frac{VC(k.s^2 + 1)}{(s^2 + A^2)(s+B)(s+D)} = VC\left(\frac{a}{(s^2 + A^2)} + \frac{b}{(s+B)} + \frac{c}{(s+D)} \right)$$ (180-8)

The new constants a, b, and c, are also determined by the same method of equting the coefficients of equal powers of s in the equation

$a (s + b)(s + D) + b(s^2 + A^2)(s + D) + c(s^2 + A^2)(s + B) = ks^2 + 1$ (180-8.1)

After executing the same steps in determining a, b, and c, we get the inverse Laplace transform

$$I(t) = VC\left(L^{-1}\left\{ \frac{a}{(s^2 + A^2)} + \frac{b}{(s+B)} + \frac{c}{(s+D)} \right\} \right) = VC\left(\frac{a}{A}\sin at + be^{-Bt} + ce^{-Dt} \right)$$ (180-9)

3.3. Nonresistive CL circuits

Example 113   **CL Networks**

In the following network, the switch S is closed at time t = o when both condensers are **charged** to voltage E. Find the current I.

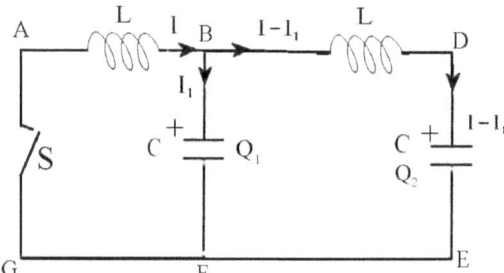

Solution

Applying the **Kirchoff's voltage law** to the circuit **ABFGA** we get

$$L\frac{dI}{dt} + \frac{Q_1}{C} = 0 \tag{181}$$

Where

$$I_1 = \frac{dQ_1}{dt} \tag{181-1}$$

Applying the voltage law to the circuit **ABDEFGA** we get

$$L\frac{dI}{dt} + L\frac{d(I-I_1)}{dt} + \frac{Q_2}{C} = 0 \tag{181-2}$$

$$I - I_1 = \frac{dQ_2}{dt} \tag{181-3}$$

Initially, the circuit possess electrostatic charge

$$Q_1(0) = Q_2(0) = E\,C \tag{181-4}$$
$$I_1(0) = I_2(0) = 0 \tag{181-5}$$

Therefore, the Laplace transforms of (181) and (181-1) are

$$L(s.i(s) - I(0)) + \frac{q_1(s)}{C} = 0 \tag{181-6.1}$$

$$i_1(s) = s.q(s) - Q(0) \tag{181-6.2}$$

214

Thus, we have the **first relation** between the two currents $i(s)$ and $i_1(s)$, as follows:

(i) $\qquad L.s.i(s) + \dfrac{i_1(s) + EC}{Cs} = 0$

$\qquad\qquad i_1(s) = -LC.s^2.i(s) - EC$ $\hfill (181\text{-}6.3)$

Similarly, the Laplace transforms of (181-2) and (181-3) are

$\qquad\qquad L.[s.i(s) - I(0)] + L.s.[i(s) - i_1(s)] + \dfrac{q(s)}{C} = 0$ $\hfill (181\text{-}7.1)$

$\qquad\qquad i(s) - i_1(s) = s.q(s) - Q(0)$ $\hfill (181\text{-}7.2)$

Substituting from (181-7.2) in (181-7.1), we get the **second relation** between the two currents $i(s)$ and $i_1(s)$, as follows:

(ii) $\qquad L.s.i(s) + L.s.[i(s) - i_1(s)] + \dfrac{i(s) - i_1(s) + EC}{Cs} = 0$ $\hfill (181\text{-}7.3)$

Eliminating $i_1(s)$ between (181-6.3) and (181-7.3) we get

$$\left(2L.s + \frac{1}{Cs}\right).i(s) - i_1(s)\left(L.s + \frac{1}{Cs}\right) + \frac{EC}{Cs} = 0$$

Where, substituting from (181-6.3), we get

$$\left(2L.s + \frac{1}{Cs}\right).i(s) + \left(LCs^2.i(s) + EC\right)\left(L.s + \frac{1}{Cs}\right) + \frac{E}{s} = 0$$

$$\left(2L.s + \frac{1}{Cs} + LCs^2\left(L.s + \frac{1}{Cs}\right)\right).i(s) + EC\left(L.s + \frac{1}{Cs}\right) + \frac{E}{s} = 0$$

$$\left(2L.Cs^2 + 1 + LCs^2\left(L.Cs^2 + 1\right)\right)i(s) + EC\left(CL.s^2 + 1\right) + EC = 0$$

$$\left(3L.Cs^2 + 1 + \left(LCs^2\right)^2\right)i(s) = -EC\left(CL.s^2 + 2\right)$$

$$i(s) = \frac{-EC\left(CL.s^2 + 2\right)}{\left(LCs^2\right)^2 + 3L.Cs^2 + 1}$$ $\hfill (181\text{-}8)$

Therefore, from (181-6.3) and (181-8) the secondary current Laplace transform is

$$i_1(s) = \frac{ELC^2s^2\left(CL.s^2 + 2\right)}{\left(LCs^2\right)^2 + 3L.Cs^2 + 1} - EC$$ $\hfill (181\text{-}9)$

The inverse Laplace transforms are obtained by putting $n^2 = 1 / CL$ in equation (181-8).

$$i(s) = \frac{-EC(CLs^2 + 2)}{3L.Cs^2 + 1 + (LCs^2)^2}$$

$$= -\left(\frac{E}{L}\right)\frac{s^2 + 2n^2}{3n^2s^2 + n^4 + s^4} \qquad (181\text{-}10)$$

The partial fractions of (181-10) are obtained similar to the previous example, specifically, equations (180-7.5) through (180-7.12).

$$i(s) = -\left(\frac{E}{2L\sqrt{5}}\right)\left(\frac{1+\sqrt{5}}{s^2 + \frac{1}{2}(3-\sqrt{5})n^2} - \frac{1-\sqrt{5}}{s^2 + \frac{1}{2}(3+\sqrt{5})n^2}\right) \qquad (181\text{-}11)$$

The inverse Laplace transform of (181-11) is obtained from the elementary functions as usual, as follows

$$I(t) = -\left(\frac{E}{2L\sqrt{5}}\right)L^{-1}\left(\frac{1+\sqrt{5}}{s^2 + \frac{1}{2}(3-\sqrt{5})n^2} - \frac{1-\sqrt{5}}{s^2 + \frac{1}{2}(3+\sqrt{5})n^2}\right)$$

$$= -E\sqrt{\frac{C}{10L}}\left[\frac{1+\sqrt{5}}{\sqrt{3-\sqrt{5}}}\sin\left(nt\sqrt{\frac{1}{2}(3-\sqrt{5})}\right) - \frac{1-\sqrt{5}}{\sqrt{3+\sqrt{5}}}\sin\left(nt\sqrt{\frac{1}{2}(3+\sqrt{5})}\right)\right] \quad (181\text{-}12)$$

3.4. RCL network circuits

Example 114   **RCL Networks**

In the network shown below, determine the character of the current $I_1(t)$ assuming that each current is zero when the switch is closed.

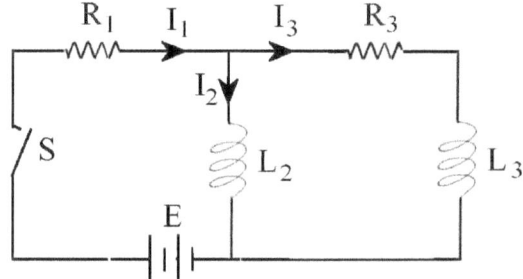

Solution

Since the algebraic sum of the currents at any junction is zero, then

$$I_1 - I_2 - I_3 = 0 \tag{182}$$

Applying the voltage law to the circuit on the left we get

$$I_1.R_1 + L_2\frac{dI_2}{dt} = E \tag{182-1}$$

Applying again the voltage low to the outside circuit we get

$$I_1.R_1 + I_3.R_3 + L_3\frac{dI_3}{dt} = E \tag{182-2}$$

Transforming (182), (182-1) and (182-2), we get

$$i_1(s) - i_2(s) - i_3(s) = 0 \tag{182-3}$$

$$R_1 i_1(s) + s.L_2 i_2(s) = \frac{E}{s} \tag{182-4}$$

$$R_1 i_1(s) + (R_3 + s.L_3)\, i_3(s) = \frac{E}{s} \tag{182-5}$$

The three linear simultaneous equations (182-3), (182-4), and (182-5) have the three unknown $i_1(s)$, $i_2(s)$, and $i_3(s)$ and can be solved by Cramer's rule of matrices among other simple methods of elimination, as follows.

$$i_1(s) = \frac{\begin{vmatrix} 0 & -1 & -1 \\ E/s & sL_2 & 0 \\ E/s & 0 & R_3 + sL_3 \end{vmatrix}}{\Delta} = \frac{E}{s} \cdot \frac{R_3 + s(L_2 + L_3)}{\Delta} \tag{182-6}$$

Where, the determinant fo the matrix is determined as follows

$$\Delta = \begin{vmatrix} 1 & -1 & -1 \\ R_1 & sL_2 & 0 \\ R_1 & 0 & R_3 + sL_3 \end{vmatrix} = \begin{vmatrix} 1 & 0 & 0 \\ R_1 & sL_2 + R_1 & R_1 \\ R_1 & R_1 & R_1 + R_3 + sL_3 \end{vmatrix}$$

$$\Delta = s^2 L_2 L_3 + s.(R_1 L_2 + R_3 L_2 + R_1 L_3) + R_1 R_3 \tag{182-6.1}$$

Since we are interested in the factors of Δ, we consider the equation $\Delta = 0$. Since all coefficients of this equation are positive, hence it cannot have any positive roots. Its discriminant is

$$\left(R_1 L_2 + R_3 L_2 + R_1 L_3\right)^2 - 4L_2 L_3 R_1 R_3$$

which can be written

$$R_1^2 L_2^2 + 2R_1 L_2\left(R_3 L_2 + R_1 L_3\right) + \left(R_3 L_2 - R_1 L_3\right)^2$$

which is positive. Hence the equation $\Delta = 0$ has two negative distinct roots $-\alpha_1$ and $-\alpha_2$, say.

Therefore,

$$\Delta = L_2 L_3 (s + \alpha_1)(s + \alpha_2) \tag{182-6.2}$$

Therefore, equations (182-6) and (186-6.2) give

$$i_1(s) = \frac{E}{s} \cdot \frac{R_3 + s(L_2 + L_3)}{L_2 L_3\left(s + \alpha_1\right)\left(s + \alpha_2\right)}$$

$$= \frac{A_0}{s} + \frac{A_1}{s + \alpha_1} + \frac{A_2}{s + \alpha_2} \tag{182-7}$$

The inverse Laplace transform of (182-7) is therefore,

$$I(t) = L^{-1}\left\{\frac{A_0}{s} + \frac{A_1}{s + \alpha_1} + \frac{A_2}{s + \alpha_2}\right\} = A_0 + A_1 e^{-\alpha_1 t} + A_2 e^{-\alpha_2 t}$$

Example 115   **RCL Networks**

In the network shown below, derive the equations satisfied by the currents I_1, I_2, I_3 and the charge Q_3 assuming that all initial currents and charges are zero and obtain the transformed equations.

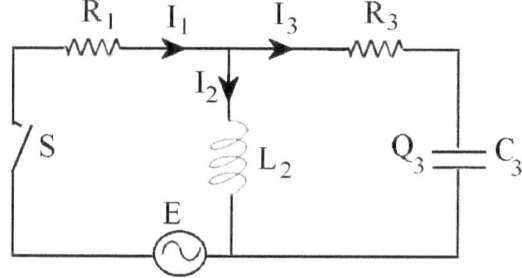

Solution

$$I_1 - I_2 - I_3 = 0 \tag{183}$$

$$I_1.R_1 + L_2 \frac{dI_2}{dt} = E\sin\omega t \tag{183-1}$$

$$I_1.R_1 + I_3.R_3 + \frac{Q_3}{C_3} = E\sin\omega t \tag{183-2}$$

$$I_3 = \frac{dQ_3}{dt} \tag{183-3}$$

Laplace Transforming equations (183), (183-1), (183-2), and (183-3) we get

$$i_1(s) - i_2(s) - i_3(s) = 0 \tag{183-4}$$

$$R_1 i_1(s) + s.L_2 i_2(s) = \frac{E\omega}{s^2 + \omega^2} \tag{183-5}$$

$$R_1 i_1(s) + R_3 i_3(s) + \frac{q_3}{C_3} = \frac{E\omega}{s^2 + \omega^2} \tag{183-6}$$

$$i_3(s) = sq_3(s) - Q_3(0) \tag{183-7}$$

Initial conditions are

$$I_1(0) = I_2(0) = I_3(0) = 0$$
$$Q_3(0) = 0$$

The solution can then be completed as in the previous example,

Example 116   Impulsive RCL Networks

An impulsive E.M.F. $E_o \delta(t)$ is applied at $t = 0$ to an L, C, R, circuit in series with zero initial currents and charges. Find the current at *any* instant t,

Solution

$$I.R + L\frac{dI}{dt} + \frac{1}{C}\int_0^t I(\tau)d\tau = E_o\delta(t) \tag{184}$$

$$\left(R + L.s + \frac{1}{Cs}\right)i(s) = E_o$$

$$i(s) = \frac{E_o}{R + L.s + \dfrac{1}{Cs}} \tag{184-1}$$

$$= \frac{E_o}{L}.\frac{s}{s^2 + \dfrac{R}{L}s + \dfrac{1}{CL}} = \frac{E_o}{L}.\frac{s}{\left(s + \dfrac{R}{2L}\right)^2 + \dfrac{1}{CL} - \left(\dfrac{R}{2L}\right)^2}$$

$$= \frac{E_o}{L} \cdot \frac{s + \dfrac{R}{2L} - \dfrac{R}{2L}}{\left(s + \dfrac{R}{2L}\right)^2 + \dfrac{1}{CL} - \left(\dfrac{R}{2L}\right)^2}$$

$$= \frac{E_o}{L}\left(\frac{s + \dfrac{R}{2L}}{\left(s + \dfrac{R}{2L}\right)^2 + \dfrac{1}{CL} - \left(\dfrac{R}{2L}\right)^2} - \frac{\dfrac{R}{2L}}{\left(s + \dfrac{R}{2L}\right)^2 + \dfrac{1}{CL} - \left(\dfrac{R}{2L}\right)^2}\right) \qquad (184\text{-}2)$$

In equation (184-20, we added and subtracted the term R / 2 L in order to render the transform function elementary and thus easy to inverse, as follows.

$$I(t) = \frac{E_o}{L}\left\{ L^{-1}\left\{ \frac{s + \dfrac{R}{2L}}{\left[\left(s + \dfrac{R}{2L}\right)^2 + \dfrac{1}{CL} - \left(\dfrac{R}{2L}\right)^2\right]} - \frac{\dfrac{R}{2L}}{\left(s + \dfrac{R}{2L}\right)^2 + \dfrac{1}{CL} - \left(\dfrac{R}{2L}\right)^2}\right\}\right\}$$

$$= \frac{E_o}{L} e^{-\frac{R}{2L}t} \cos nt - \frac{E_o}{L}\left(\frac{R}{2L}\right)\frac{1}{n} e^{-\frac{R}{2L}t} \sin nt \qquad (184\text{-}3)$$

Where, $\qquad n^2 = \dfrac{1}{CL} - \left(\dfrac{R}{2L}\right)^2 > 0$

CHAPTER 4

DYNAMICAL APPLICATIONS OF LAPLACE TRANSFORMS

Example 117 Displacement and motion under gravity and resistance

A particle is projected vertically upwards at time t = 0 with **velocity** v_0 from the origin under the action of **gravity** and a **resistance** equal to 2 km times the velocity. Find its **displacement** at any instant t.

Solution

The **equation of motion** of the mass is the balance of induced forces (mass x acceleration) against acting forces (gravity + resistance, both directed downward)

$$m\ddot{x} = -2km\dot{x} - mg \qquad (185)$$

with the initial conditions

$$x(0) = 0 \qquad \dot{x}(0) = v_0 \qquad (185\text{-}1)$$

The Laplace transform of (185),after division by mass m of both sides,is

$$s^2 x(s) - s.x(0) - \dot{x}(0) = -2k.[s.x(s) - x(0)] - \frac{g}{s} \qquad (185\text{-}2)$$

From the initial conditions in (185-1), equation (165-2) becomes

$$s^2 x(s) - v_0 = -2k.s.x(s) - \frac{g}{s}$$

Therefore,

$$x(s)\left(s^2 + 2k\right) = v_0 - \frac{g}{s}$$

$$x(s) = \frac{v_0 - \frac{g}{s}}{s^2 + 2k} = \frac{v_0}{s^2 + 2k} - \frac{g}{s\left(s^2 + 2k\right)} \qquad (185\text{-}3)$$

The partial fractions of (185-3) take the following form

$$x(s) = \frac{v_0 / 2k}{s} - \frac{v_0 / 2k}{s + 2k} - \frac{g / 2k}{s^2} + \frac{g /(2k)^2}{s} - \frac{g /(2k)^2}{s + 2k} \qquad (185\text{-}4)$$

We have skipped the steps of determining the constants of the partial fractions from the coefficients of equal powers in s, since we have done so extensively before, in this book.

The inverse Laplace transform of (185-4) is

$$x(t) = L^{-1}\left\{\frac{v_o/2k}{s} - \frac{v_o/2k}{s+2k} - \frac{g/2k}{s^2} + \frac{g/(2k)^2}{s} - \frac{g/(2k)^2}{s+2k}\right\}$$

$$= \frac{v_o}{2k} - \frac{v_o}{2k}e^{-2kt} - \frac{g}{2k}t + \frac{g}{(2k)^2} - \frac{g}{(2k)^2}e^{-2kt}$$

$$= \frac{v_o}{2k}\left(1 - e^{-2kt}\right) - \frac{g}{2k}t + \frac{g}{(2k)^2}\left(1 - e^{-2kt}\right)$$

$$= -\frac{gt}{2k} + \frac{g+2kv_o}{(2k)^2}\left(1 - e^{-2kt}\right) \qquad\qquad (185\text{-}5)$$

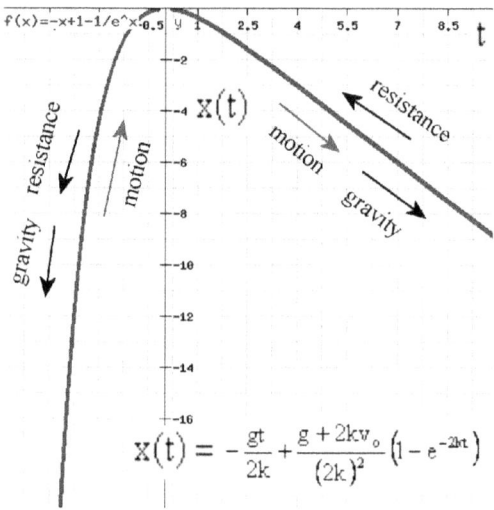

$$x(t) = -\frac{gt}{2k} + \frac{g+2kv_o}{(2k)^2}\left(1 - e^{-2kt}\right)$$

Example 118 Motion under central force

A particle of **mass** m moves in a vertical plane under the action of **gravity** and **a force directed** towards the origin and equal to μr where r is the distance of the particle from the origin. If the particle is projected from the point (a, 0) vertically upwards with velocity v_o, find the coordinates of the particle at any instant t.

Solution

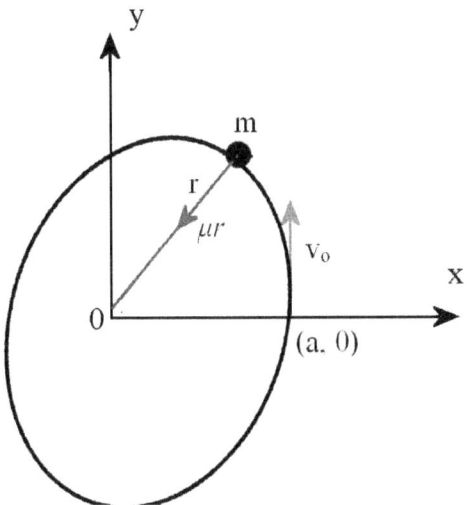

The equations of motion under central force μx is given by equating the induced acceleration inertia by the dragging force. Of course, the mass must be energized in order to move, hence the initial velocity v_0.

Horizontal forces $\qquad\qquad m\ddot{x} = -\mu x$ $\qquad\qquad\qquad\qquad\qquad$ (186)

Vertical forces $\qquad\qquad m\ddot{y} = -\mu y - mg$ $\qquad\qquad\qquad\qquad$ (186-1)

Divide both equations by m and substitute by, $n^2 = \mu/m$, to get

Horizontal forces $\qquad\qquad \ddot{x} = -n^2 x$ $\qquad\qquad\qquad\qquad\qquad$ (186-2)

Vertical forces $\qquad\qquad \ddot{y} = -n^2 y - g$ $\qquad\qquad\qquad\qquad\quad$ (186-3)

The initial conditions are

$$x(0) = 0$$
$$y(0) = 0$$
$$\dot{x}(0) = 0$$
$$\dot{y}(0) = v_0$$

Thus, the Laplace transforms of equations (186-2) and (186-3) are

$$s^2.x(s) - s.x(0) - \dot{x}(0) = -n^2 x(s)$$

The initial conditions give

$$x(s) = \frac{s.a}{s^2 + n^2} \qquad (186\text{-}4)$$

On inversing, the temporal profile x(t) is given by

$$x(t) = L^{-1}\left\{\frac{s.a}{s^2 + n^2}\right\} = a\cos nt \qquad (186\text{-}4.1)$$

Similarly, the Laplace transform of (186-3) is

$$s^2.y(s) - s.y(0) - \overset{..}{y}(0) = -n^2 x(s) - \frac{g}{s} \qquad (186\text{-}5)$$

Substituting by the initial conditions in (186-5) we get

$$s^2.y(s) - v_0 = -n^2 y(s) - \frac{g}{s}$$

$$y(s) = \frac{v_0 - \dfrac{g}{s}}{s^2 + n^2}$$

$$= \frac{v_0}{s^2 + n^2} - \frac{g}{s(s^2 + n^2)} \qquad (186\text{-}6)$$

On inversing, the temporal profile y(t) is given by

$$y(t) = L^{-1}\left\{\frac{v_0}{s^2 + n^2} - \frac{g}{s(s^2 + n^2)}\right\} = \frac{v_0}{n}\sin nt - \frac{g}{n^2}(1 - \cos nt) \qquad (186\text{-}7)$$

Example 119   Motion under varying forces

A particle of mass m moves in a straight line under a resistance 2 μm times the velocity and a restoring force **mλ** times the displacement where $\lambda > \mu^2$. If it is projected at time t = 0, with velocity v_0 at distance x_0; from its equilibrium position, find its displacement at any subsequent instant t.

Solution

Equation of motion of the mass is

$$m\overset{...}{x} = -2\mu m\,\overset{..}{x} - m\lambda x$$

i.e., $$\overset{...}{x} + 2\mu\,\overset{..}{x} + \lambda x = 0 \qquad (187)$$

with the initial conditions

$$x(0) = 0 \text{ and } \overset{..}{x}(x_0) = v_0$$

The Laplace transform of (187) is

$$s^2 x(s) - s x(0) - \overset{\bullet}{x}(0) + 2\mu[s x(s) - x(0)] + \lambda x(s) = 0$$

$$s^2 x(s) - s x_o - v_o + 2\mu[s x(s) - x_o] + \lambda x(s) = 0$$

$$x(s)\left(s^2 + 2\mu s + \lambda\right) = s x_o + v_o + 2\mu x_o$$

$$x(s) = \frac{s x_o + v_o + 2\mu x_o}{s^2 + 2\mu s + \lambda} = \frac{s x_o + v_o + 2\mu x_o}{\left(s + \mu\right)^2 + \lambda - \mu^2} \tag{187-1}$$

Put $n^2 = \lambda - \mu^2$

Therefore, (187-1) becomes

$$x(s) = \frac{s x_o + v_o + 2\mu x_o}{s^2 + 2\mu s + \lambda} = \frac{x_o(s + \mu) + v_o + \mu x_o}{\left(s + \mu\right)^2 + n^2}$$

$$= \frac{x_o(s + \mu)}{\left(s + \mu\right)^2 + n^2} + \frac{v_o + \mu x_o}{\left(s + \mu\right)^2 + n^2} \tag{187-2}$$

The inverse Laplace transform of (187-2) is

$$x(t) = L^{-1}\left\{\frac{x_o(s + \mu)}{\left(s + \mu\right)^2 + n^2} + \frac{v_o + \mu x_o}{\left(s + \mu\right)^2 + n^2}\right\}$$

$$= x_o e^{-\mu t} \cos nt + \frac{v_o + \mu x_o}{n} e^{-\mu t} \sin nt$$

$$= \frac{e^{-\mu t}}{n}\left(n x_o \cos nt + \left(v_o + \mu x_o\right)\sin nt\right) \tag{187-3}$$

Example 120 Elastic spring motion

A **spring** of stiffness k is placed on a smooth horizontal table. One end of the spring is fixed at a point O on the table and to the other end is attached a **mass** m. The system is initially at rest with the spring not stretched .A **constant force** F is applied to the mass for a time t_o and then removed. Find the mass at any time t.

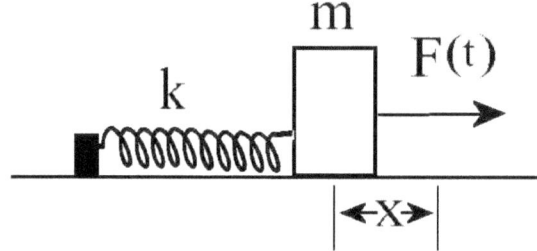

Equation of motion of the mass is

$$m\ddot{x} = -kx + F(t)$$
$$\ddot{x} = -\frac{k}{m}x + \frac{F(t)}{m} \qquad (188)$$

Put $\qquad n^2 = \dfrac{k}{m}$

Thus, equation (188) becomes

$$\ddot{x} + n^2 x = \frac{F(t)}{m} \qquad (188\text{-}1)$$

Where $\qquad F(t) = F_o \qquad\qquad 0 < t < t_o$
$$= 0 \qquad\qquad\quad t > t_o$$

The initial conditions are

$$x(0) = \ddot{x}(0) = 0$$

Laplace transforming of (188-1) and noticing step transform

$$L\left\{ F(t) \right\} = \frac{F_o}{s}\left(1 - e^{-t_o s}\right) \qquad (188\text{-}2)$$

The Laplace transform of (188-1) becomes

$$\left(s^2 + n^2\right)x(s) = \frac{F_o}{ms}\left(1 - e^{-t_o s}\right)$$

$$x(s) = \frac{F_o}{ms\left(s^2 + n^2\right)}\left(1 - e^{-t_o s}\right)$$

$$x(s) = \frac{F_o}{mn^2}\left(\frac{1}{s} - \frac{s}{s^2 + n^2}\right)\left(1 - e^{-t_o s}\right)$$ (188-3)

The inverse Laplace transform of (188-3) is

$$x(t) = \frac{F_o}{mn^2}L^{-1}\left\{\left(\frac{1}{s} - \frac{s}{s^2 + n^2}\right)\left(1 - e^{-t_o s}\right)\right\}$$

$$= \frac{F_o}{mn^2}L^{-1}\left\{\frac{1}{s} - \frac{s}{s^2 + n^2} - \left(\frac{1}{s} - \frac{s}{s^2 + n^2}\right)e^{-t_o s}\right\}$$

$$= \frac{F_o}{mn^2}\left(1 - \cos nt - U(t - t_o)(1 - \cos[n(t - t_o)])\right)$$ (188-4)

******** **At** $0 < t < t_o$, $U(t - t_o) = 0$

Therefore, (188-4) becomes

$$x(t) = \frac{F_o}{mn^2}(1 - \cos nt)$$ (188-5)

Since, $\cos nt = \cos^2(nt/2) - \sin^2(nt/2) = 1 - 2\sin^2(nt/2)$ (188-5.1)

Therefore, (188-5) becomes

$$x(t) = \frac{2F_o}{mn^2}\sin^2\left(\frac{nt}{2}\right)$$ (188-5.2)

******** **At** $t > t_o$, $U(t - t_o) = 1$

Therefore, (188-4) becomes

$$x(t) = \frac{F_o}{mn^2}\left(\cos[n(t - t_o)] - \cos nt\right)$$ (188-6)

Also,

$$\cos n(t-t_o) = \cos^2(n(t-t_o)/2) - \sin^2(n(t-t_o)/2) = 1 - 2\sin^2(n(t-t_o)/2)$$ (188-6.1)

Therefore, (188-6) becomes

$$x(t) = \frac{2F_o}{mn^2}\left(\sin^2\left(\frac{nt}{2}\right) - \sin^2\left(\frac{n(t - t_o)}{2}\right)\right)$$ (188-6.2)

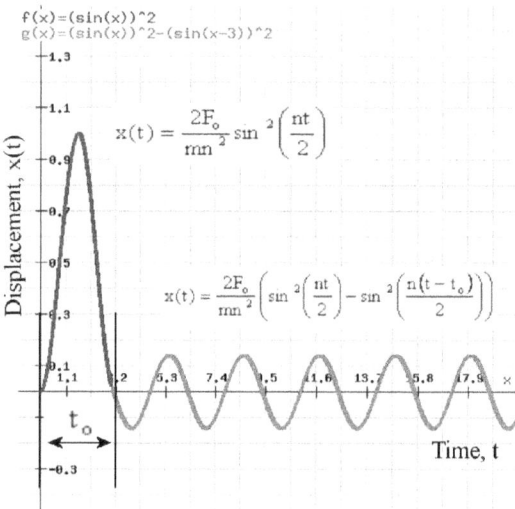

f(x)=(sin(x))^2
g(x)=(sin(x))^2-(sin(x-3))^2

$$x(t) = \frac{2F_0}{mn^2}\sin^2\left(\frac{nt}{2}\right)$$

$$x(t) = \frac{2F_0}{mn^2}\left(\sin^2\left(\frac{nt}{2}\right) - \sin^2\left(\frac{n(t-t_0)}{2}\right)\right)$$

Displacement, x(t)

Time, t

Example 121 Elastic spring motion

Two particles each of **mass m** are connected by a spring of **stiffness k** and they are free to move in a straight line on a smooth horizontal table. At time t = o when both particles are at rest and the spring is unstrained a **constant force P** is applied to one of them in the direction towards the other particle. Find the displacement of the other particle from its initial position at any subsequent instant t.

Solution

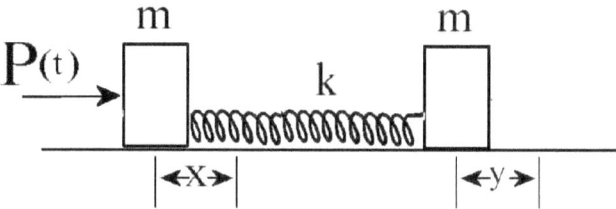

The equations of motion of the two masses are

$$m\,\overset{\cdot\cdot}{x}(t) = k\left[y(t) - x(t)\right] + P \qquad (189)$$

$$m\,\overset{\cdot\cdot}{y}(t) = -k\left[y(t) - x(t)\right] \qquad (189\text{-}1)$$

Putting $k/m = n^2$ we get

$$\overset{\cdot\cdot}{x}(t) = n^2\left[y(t) - x(t)\right] + \frac{P}{m} \qquad (189\text{-}2)$$

$$\overset{\cdot\cdot}{y}(t) = -n^2\left[y(t) - x(t)\right] \qquad (189\text{-}3)$$

The initial conditions are

$$x(0) = 0$$
$$y(0) = 0$$
$$\overset{\cdot\cdot}{x}(0) = 0$$
$$\overset{\cdot}{y}(0) = 0$$

The Laplace transforms of (189-2) and (189-3) is

$$s^2 x(s) = n^2\left[y(s) - x(s)\right] + \frac{P}{ms} \qquad (189\text{-}4)$$

$$\left(s^2 + n^2\right)x(s) = n^2 y(s) + \frac{P}{ms}$$

$$x(s) = \frac{n^2 m s . y(s) + P}{ms\left(s^2 + n^2\right)} \qquad (189\text{-}4.1)$$

$$s^2 y(s) = -n^2\left[y(s) - x(s)\right]$$

$$y(s) = \frac{n^2}{\left(s^2 + n^2\right)} x(s) \qquad (189\text{-}5)$$

Eliminate x(s) between (189-4.1) and (189-5) we get

$$y(s) = \frac{n^2}{\left(s^2 + n^2\right)} \cdot \frac{n^2 m s . y(s) + P}{ms\left(s^2 + n^2\right)}$$

$$\left[\left(s^2 + n^2\right)^2 - n^4\right] y(s) = \frac{n^2 P}{ms} \qquad (189\text{-}6)$$

The square-difference reduces to

$$(s^2 + n^2 - n^2)(s^2 + n^2 + n^2) = s^2(s^2 + 2n^2) \qquad (189\text{-}7)$$

229

Thus, equation (189-6) gives

$$y(s) = \frac{n^2 P}{ms} \cdot \frac{1}{s^2 (s^2 + 2n^2)}$$

$$= \frac{n^2 P}{4m} \left(\frac{2}{s^3} - \frac{1}{n^2 s} + \frac{s}{n^2 (s^2 + 2n^2)} \right) \qquad (189-8)$$

The partial fraction form in (189-8) is obtained in the same fashion by Heaviside method, section 2.21.

The Laplace inversion of (189-8) is

$$y(t) = \frac{n^2 P}{4m} . L^{-1} \left\{ \frac{2}{s^3} - \frac{1}{n^2 s} + \frac{s}{n^2 (s^2 + 2n^2)} \right\}$$

$$= \frac{P}{4m} \left(t^2 - \frac{m}{k} + \frac{m}{k} \cos \left(\left(\sqrt{\frac{2k}{m}} \right) t \right) \right) \qquad (189-9)$$

Example 122   Two elastic springs in motion

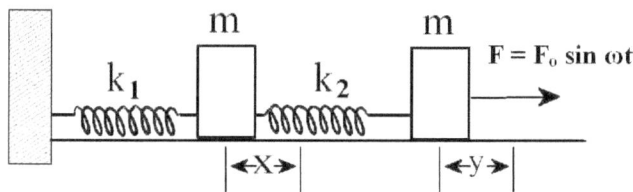

Two masses are equal each being equal to m. The stiffnesses k_1, k_2 of the two springs are connected by the relation $k_1 / k_2 = 3 / 2$. The system is initially at rest with the two springs unstrained. A periodic force $F = F_o \sin \omega t$, where $\omega^2 = k_2 / m$ is applied to the right mass. Find the displacement X(t) of the left mass at any instant t.

Solution

Equations of motion of the masses are

$$m \ddot{X}(t) = k_2 [Y(t) - X(t)] - k_1 X(t) \qquad (190)$$

$$m \ddot{Y}(t) = -k_2 [Y(t) - X(t)] + F_o \sin \omega t \qquad (190-1)$$

Substituting by $\omega^2 = k_2/m$ and $k_1/k_2 = 3/2$, we get

$$\overset{\cdots}{X}(t) = \omega^2[Y(t) - X(t)] - \frac{3}{2}\omega^2 X(t) \tag{190-2}$$

$$\overset{\cdots}{Y}(t) = -\omega^2[Y(t) - X(t)] + \frac{F_o}{m}\sin \omega t \tag{190-3}$$

Initial conditions are

$$X(0) = 0$$
$$Y(0) = 0$$
$$\overset{..}{X}(0) = 0$$
$$\overset{..}{Y}(0) = 0$$

The Laplace transforms of (190-2) and (190-3) are

$$s^2 x(s) = \omega^2\left[y(s) - \frac{5}{2}x(s)\right]$$

$$x(s) = \frac{\omega^2}{s^2 + \frac{5}{2}\omega^2}\, y(s) \tag{190-4}$$

And

$$y(s) = \frac{\omega^2}{s^2 + \omega^2}\, x(s) + \frac{F_o}{m}\,\frac{\omega}{\left(s^2 + \omega^2\right)^2} \tag{190-5}$$

Eliminating $y(s)$ between (190-4) and (190-5) we get

$$x(s)\frac{\left(s^2 + \frac{5}{2}\omega^2\right)}{\omega^2} = \frac{\omega^2}{s^2 + \omega^2}\, x(s) + \frac{F_o}{m}\,\frac{\omega}{\left(s^2 + \omega^2\right)^2}$$

$$\left(\left(s^2 + \frac{5}{2}\omega^2\right)\left(s^2 + \omega^2\right) - \omega^4\right) x(s) = \frac{\omega^3 F_o}{m}\,\frac{1}{s^2 + \omega^2} \tag{190-6}$$

$$\left(s^4 + \frac{7}{2}s^2\omega^2 + \frac{3}{2}\omega^4\right) x(s) = \frac{\omega^3 F_o}{m}\,\frac{1}{s^2 + \omega^2}$$

$$\left(s^2 + \frac{1}{2}\omega^2\right)\left(s^2 + 3\omega^2\right) x(s) = \frac{\omega^3 F_o}{m}\,\frac{1}{s^2 + \omega^2}$$

231

$$x(s) = \frac{\omega^3 F_o}{m} \frac{1}{\left(s^2 + \omega^2\right)\left(s^2 + \frac{1}{2}\omega^2\right)\left(s^2 + 3\omega^2\right)}$$

$$= \frac{\omega^5 F_o}{k_2}\left[\frac{4}{5\omega^4\left(s^2 + \frac{1}{2}\omega^2\right)} + \frac{1}{5\omega^4\left(s^2 + 3\omega^2\right)} - \frac{1}{\omega^4\left(s^2 + \omega^2\right)}\right] \qquad (190\text{-}7)$$

The inverse Laplace transform of (190-7) is

$$X(t) = \frac{\omega F_o}{5k_2} L^{-1}\left\{\frac{4}{\left(s^2 + \frac{1}{2}\omega^2\right)} + \frac{1}{\left(s^2 + 3\omega^2\right)} - \frac{5}{\left(s^2 + \omega^2\right)}\right\}$$

$$X(t) = \frac{\omega F_o}{5k_2}\left\{\frac{4\sqrt{2}}{\omega}\sin\left(\sqrt{\frac{1}{2}}\omega t\right) + \frac{1}{\omega\sqrt{3}}\sin\left(\sqrt{3}\omega t\right) - \frac{5}{\omega}\sin(\omega t)\right\} \qquad (190\text{-}8)$$

Example 123 Angular inertial motion with couples

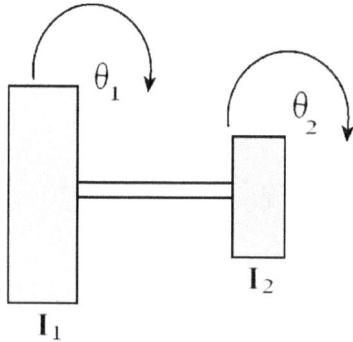

Two flywheels of **moments of inertia** I_1 and I_2 are connected by an elastic shaft of torsional stiffness A i.e., the **couple per radian** relative twist of the flywheels is A. The whole system is rotating with a **constant angular velocity** ω when at time $t = 0$, a constant **retarding couple** P is applied to the wheel I_1.

232

(i) Find the angular velocity of the wheel I_2 at any instant t.

(ii) Let the retarding couple P be applied for time T only and find the angular velocity of the wheel I_2 at any instant t.

Solution

(i) P be applied for time t = 0

Let θ_1 be the angular displacement of the flywheel I_1 and θ_2 that of I_2.

Equations of motion are

$$I_1 \, \ddot{\theta}_1(t) = \lambda\big[\theta_2(t) - \theta_1(t)\big] - P \tag{191}$$

$$I_2 \, \ddot{\theta}_2(t) = -\lambda\big[\theta_2(t) - \theta_1(t)\big] \tag{191-1}$$

Initial conditions can be taken as

$$\theta_1(0) = 0$$
$$\theta_2(0) = 0$$
$$\dot{\theta}_1(0) = \omega$$
$$\dot{\theta}_2(0) = \omega$$

Therefore, the Laplace transforms of (190) and (190-1) become

$$I_1[s^2\theta_1(s) - s\theta_1(0) - \dot{\theta}_1(0)] = \lambda[\theta_2(s) - \theta_1(s)] - \frac{P}{s} \tag{191-2}$$

$$I_2[s^2\theta_2(s) - s\theta_2(0) - \dot{\theta}_2(0)] = -\lambda[\theta_2(s) - \theta_1(s)] \tag{191-3}$$

Upon substituting by the initial conditions we get

$$I_1[s^2\theta_1(s) - \omega] = \lambda[\theta_2(s) - \theta_1(s)] - \frac{P}{s} \tag{191-4}$$

$$I_2[s^2\theta_2(s) - \omega] = -\lambda[\theta_2(s) - \theta_1(s)] \tag{191-5}$$

Arranging the terms in (191-4) and (191 5), we get

$$\theta_1(s) = \frac{\lambda}{I_1 s^2 + \lambda}\theta_2(s) + \frac{I_1\omega s - \dfrac{P}{s}}{I_1 s^2 + \lambda} \tag{191-6}$$

233

$$\theta_2(s) = \frac{I_2\omega}{I_2s^2 + \lambda} + \frac{\lambda}{I_2s^2 + \lambda}\theta_1(s) \qquad (191\text{-}7)$$

Eliminate $\theta_1(s)$ between (191-6) and (191-7)

$$\theta_2(s) = \frac{I_2\omega}{I_2s^2 + \lambda} + \frac{\lambda}{I_2s^2 + \lambda}\left(\frac{\lambda}{I_1s^2 + \lambda}\theta_2(s) + \frac{I_1\omega s - \dfrac{P}{s}}{I_1s^2 + \lambda}\right)$$

$$\theta_2(s)\left(1 - \frac{\lambda}{I_1s^2 + \lambda}\cdot\frac{\lambda}{I_2s^2 + \lambda}\right) = \frac{I_2\omega}{I_2s^2 + \lambda} + \frac{\lambda}{I_2s^2 + \lambda}\cdot\frac{I_1\omega - \dfrac{P}{s}}{I_1s^2 + \lambda}$$

$$\theta_2(s)\left(\left(I_1s^2 + \lambda\right)\left(I_2s^2 + \lambda\right) - \lambda^2\right) = I_2\omega\left(I_1s^2 + \lambda\right) + \lambda\left(I_1\omega - \frac{P}{s}\right)$$

$$\theta_2(s)s^2\left[I_1I_2s^2 + \lambda\left(I_1 + I_2\right)\right] = \omega\left[I_1I_2s^2 + \lambda\left(I_1 + I_2\right)\right] - \frac{\lambda P}{s}$$

$$\theta_2(s) = \frac{\omega}{s^2} - \frac{\lambda P}{s^3\left[I_1I_2s^2 + \lambda\left(I_1 + I_2\right)\right]} \qquad (191\text{-}8)$$

Let

$$\varphi(t) = \ddot{\theta}_2(t) \qquad (191\text{-}9)$$

The Laplace transform of (191-9) is

$$\varphi(s) = s\theta_2(s) - \theta_2(0) \qquad (191\text{-}10)$$

From (191-8) and (191-10), we get

$$\phi(s) = \frac{\omega}{s} - \frac{\lambda P}{s^2\left[I_1I_2s^2 + \lambda\left(I_1 + I_2\right)\right]} \qquad (191\text{-}11)$$

Substitute by

$$n^2 = \frac{\lambda\left(I_1 + I_2\right)}{I_1I_2} \qquad (191\text{-}12)$$

From (191-11) and (191-12), we can write

$$\phi(s) = \frac{\omega}{s} - \frac{\lambda P}{I_1I_2n^2}\left[\frac{1}{s^2} - \frac{1}{s^2 + n^2}\right] \qquad (191\text{-}13)$$

Which, with the help of (191-12), can be further rewritten as

$$\phi(s) = \frac{\omega}{s} - \frac{P}{I_1 + I_2}\left[\frac{1}{s^2} - \frac{1}{s^2 + n^2}\right] \qquad (191\text{-}14)$$

The inverse Laplace transform of (191-14) is

$$\phi(t) = L^{-1}\left\{\frac{\omega}{s} - \frac{P}{I_1 + I_2}\left[\frac{1}{s^2} - \frac{1}{s^2 + n^2}\right]\right\} = \omega - \frac{Pt}{I_1 + I_2} + \frac{P}{n(I_1 + I_2)}\sin nt \qquad (191\text{-}15)$$

(ii) P be applied for time T

Solution

Here the Laplace transform of P in equation (191-2) changes from

$$\frac{P}{s} \quad \text{to} \quad \frac{P}{s}\left(1 - e^{-Ts}\right)$$

Thus, equation (191-14) becomes

$$\phi(s) = \frac{\omega}{s} - \frac{P}{I_1 + I_2}\left[\frac{1}{s^2} - \frac{1}{s^2 + n^2}\right]\left(1 - e^{-Ts}\right) \qquad (191\text{-}16)$$

Hence, the inverse Laplace Transform of (191-16) yields

$$\phi(t) = \omega - \frac{Pt}{I_1 + I_2} + \frac{P}{n(I_1 + I_2)}\sin nt + U(t - T)\left(-\frac{P(t - T)}{I_1 + I_2} + \frac{P}{n(I_1 + I_2)}\sin n(t - T)\right) \qquad (191\text{-}17)$$

Where,
$$U(t\text{-}T) = 0 \text{ where } 0 < t < T \text{ and}$$
$$U(t\text{-}T) = 1 \text{ where } t > T$$

Example 124   Tautochrone with constant time of descent

Find the displacement equation of a **tautochrone** which is the shape taken by a smooth wire through the origin in the vertical x-y plane, such that, when a bead is constrained to slide on it, **time of descent** of the bead from any point on the wire at which it is at rest to the origin, which is its lowest point, is constant. i.e., independent of the starting point.

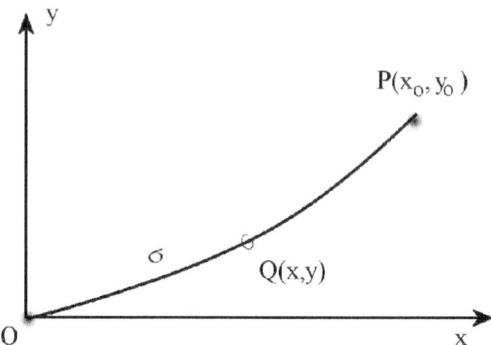

Solution

Let $P(x_0, y_0)$ be the starting print i.e., the bead is at rest at P and starts sliding and let $Q(x,y)$ be any point on the wire between O and P where O is the origin taken at the lowest point of the wire. Let $\sigma = $ arc OQ, measured from O and $m = $ mass of bead.

The loss of potential energy when the bead falls through the vertical distance y_0 - y is equal to the gain of the kinetic energy. Therefore,

$$\frac{1}{2}m\left(\frac{d\sigma}{dt}\right)^2 = mg(y_0 - y) \tag{192}$$

Thus

$$\frac{d\sigma}{dt} = -\sqrt{2g(y_0 - y)} \tag{192-1}$$

The negative sign is taken since σ decreases as t increases. Let T be the time of descent from P to O, then

$$T = -\int_{y_0}^{0} \frac{d\sigma}{\sqrt{2g(y_0 - y)}} \tag{192-2}$$

Now, when the equation of the curve is known, we can express the length of the arc σ in terms of y and consequently the differential of the arc $d\sigma$ can be expressed in terms of y_0 and dy — Hence we can put

$$d\sigma = F(y)\, dy \tag{192-3}$$

$$T\sqrt{2g} = \int_{0}^{y_0} \frac{F(y)dy}{\sqrt{y_0 - y}} \tag{192-4}$$

236

This is an integral equation of the **convolution type** in which the unknown function F(y) has to be determined such that T is constant that is independent of y_0. Now the integral equation can be written

$$T\sqrt{2g} = F(y) * y^{-\frac{1}{2}} \qquad (192\text{-}5)$$

Transforming, we get

$$\frac{T\sqrt{2g}}{s} = f(s)\sqrt{\frac{\pi}{s}}$$

$$f(s) = T\sqrt{\frac{2g}{\pi}} \cdot \frac{1}{\sqrt{s}} \qquad (192\text{-}6)$$

Inverting we get

$$F(y) = T\sqrt{\frac{2g}{\pi}} \cdot L^{-1}\left(\frac{1}{\sqrt{s}}\right) = T\sqrt{\frac{2g}{\pi}} \cdot \frac{1}{\sqrt{\pi y}} = \frac{T}{\pi}\sqrt{\frac{2g}{y}} \qquad (192\text{-}7)$$

From (192-3), we get

$$F(y) = \frac{d\sigma}{dy} = \frac{\sqrt{(dx)^2 + (dy)^2}}{dy} = \sqrt{1 + \left(\frac{dx}{dy}\right)^2} = \frac{T}{\pi}\sqrt{\frac{2g}{y}} \qquad (192\text{-}8)$$

Squaring both sides of (192-8), we get

$$1 + \left(\frac{dx}{dy}\right)^2 = \frac{2g}{y}\left(\frac{T}{\pi}\right)^2 \qquad (192\text{-}9)$$

Equation (192-9) provides the derivative of the displacement equation of the **tautochrone** wire

$$dx = \left(\sqrt{\frac{2g}{y}\left(\frac{T}{\pi}\right)^2 - 1}\right)dy \qquad (192\text{-}10)$$

Put

$$c = 2g\left(\frac{T}{\pi}\right)^2 \qquad (192\text{-}10.1)$$

Thus, (192-10) gives

$$x = \int \sqrt{\frac{c-y}{y}}dy + C \qquad (192\text{-}11)$$

Again, put
$$y = c \sin^2 \theta \qquad (192\text{-}11.1)$$

Therefore, (192-11) gives

$$x = \int \sqrt{\frac{c\ \cos^2 \theta}{c\ \sin^2 \theta}} ..2c\ \sin \theta \cos \theta\, d\theta + C$$

$$= 2c \int \cos^2 \theta\, d\theta + C$$

$$= c \int (1 + \cos 2\theta)\, d\theta + C$$

$$= \frac{c}{2}\left[2\theta + \sin 2\theta\right] + C \qquad (192\text{-}11.2)$$

Also, (192-11.1) can be written as

$$y = \frac{c}{2}\left[1 - \cos 2\theta\right] \qquad (192\text{-}11.3)$$

Now, since the curve passes through origin, therefore C = 0, and the **parametric equations** of the curve are defined by (192-10.1), (192-11.2), and (192-11.3) as follows

$$x = g\left(\frac{T}{\pi}\right)^2 \left[2\theta + \sin 2\theta\right] \qquad (192\text{-}12)$$

$$y = g\left(\frac{T}{\pi}\right)^2 \left[1 - \cos 2\theta\right] \qquad (192\text{-}13)$$

These are the parametric equations of a **cycloid**.

Hence the wire assumes the shape of a cycloid in which the radius of the generating circle is

$$g\left(\frac{T}{\pi}\right)^2$$

Example 125   Spring - mass –cam system

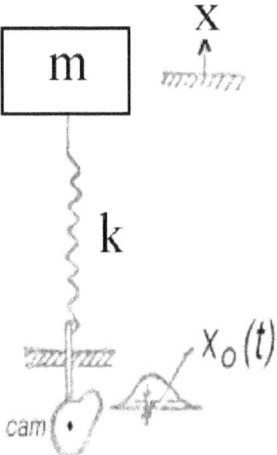

Find the **equation displacement** of a spring-mass-cam shown in the figure.
The lower end of of the spring undergoes a motion $x_0(t)$ prescribed by a cam.

Solution

The equation of motion of the mass m is

$$m\ddot{x} = k[x_0(t) - x(t)]$$

$$\ddot{x} = \frac{k}{m}[x_0(t) - x(t)] \qquad (193)$$

Put $\qquad n^2 = \frac{k}{m}$

$$\ddot{x} + n^2 x(t) = n^2 x_0(t) \qquad (193\text{-}1)$$

If the system is started from rest, the subsidiary equation for the displacement of m is the Laplace transform of (193-1) is

$$\left(s^2 + n^2\right)x(s) = n^2 x_0(s) \qquad (193\text{-}2)$$

i.e.,

$$x(s) = \frac{n^2 x_0(s)}{s^2 + n^2} \qquad (193\text{-}3)$$

When $x_0(t)$ is specified.

$x_o(s)$ can be determined and consequently $x(s)$ and $x(t)$.

To obtain the force exerted on in by the spring in which case we are concerned with the relative motion $y = x_o(t) - x$ of the ends of the spring, we have

$$F = ky = m\ddot{x}$$ (193-4)

The subsidiary equation of (193-4) is

$$F(s) = ms^2 x(s)$$ (193-5)

From (193-3) and (193-5), we get

$$F(s) = ms^2 \frac{n^2 x_o(s)}{s^2 + n^2}$$ (193-6)

Example 126   Impulse perpendicular to rod

A uniform rod of length 2a and mass m is at rest on a smooth horizontal table, At time $t = 0$, it is set in motion by a blow of **impulse** P at one end of the rod perpendicular to the rod. Find and solve its **equation of motion**.

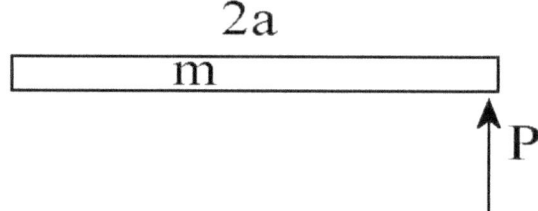

Solution

Let x be the linear displacement of the middle point of the rod and θ the angular displacement of the rod about its centre.

The moments of inertia of a rod of mass m and length 2a around its center is

$$I = \frac{mL^2}{12} = \frac{m(2a)^2}{12} = \frac{ma^2}{3}$$ (194)

Displacement: $m\ddot{x} = P\delta(t)$ (194-1)

Rotation: $\frac{1}{3}ma^2\ddot{\theta} = Pa\delta(t)$ (194-2)

Transforming (194-1) and (194-2), we get

$$ms^2x(s) = P \qquad (194\text{-}3)$$

$$\frac{1}{3}mas^2\theta(s) = P \qquad (194\text{-}4)$$

Therefore,

$$x(s) = \frac{P}{ms^2} \qquad (194\text{-}5)$$

$$\theta(s) = \frac{3P}{mas^2} \qquad (194\text{-}4)$$

CHAPTER 5

STRUCTURAL APPLICATIONS

5.1. Deflection of beams

The differential equation for the **deflection y of a beam** is

$$E.I\frac{d^2y}{dx^2} = -M \tag{195}$$

Where,

M   is the **bending moment** at a point x of the beam,
E   is its **Young's modulus** and
I   the **moment of inertia** of the cross section of the beam about its neutral axis.

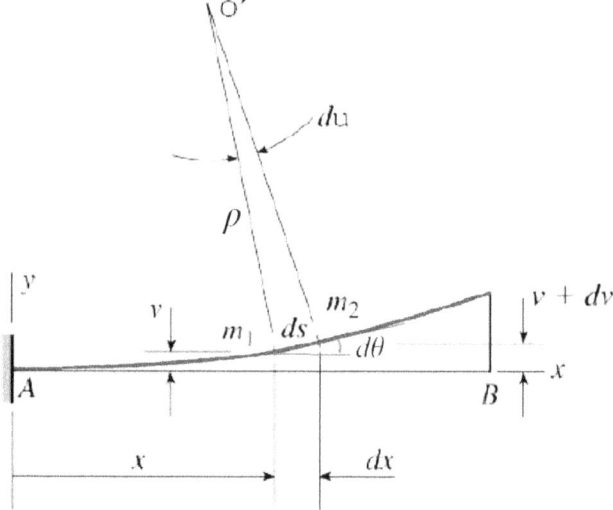

The **deflection v** is the displacement in the y direction. The **angle of rotation θ** of the axis (also called slope) is the angle between the x axis and the **tangent to the deflection** curve. **ρ is the radius of curvature**, then

$$\rho \, d\theta = ds \tag{195-1}$$

and the **curvature κ** is given by

$$\kappa = \frac{1}{\rho} = \frac{d\theta}{ds} \tag{195-2}$$

The **deflection** θ and **curvature** κ is obtained as follows

$$\frac{dv}{dx} = \tan\theta \tag{195-3}$$

As deflection is assumed small at the local level, then $\tan\theta \approx \theta$ and $ds = dx$, thus

$$\kappa = \frac{d\theta}{ds} = \frac{d}{dx}(\tan\theta) = \frac{d}{dx}\left(\frac{dv}{dx}\right) = \frac{d^2v}{dx^2} \tag{195-4}$$

For beam materials of linear elastic description, we get

$$\kappa = \frac{M}{EI} = \frac{d^2v}{dx^2} = \frac{d\theta}{dx} \tag{195-5}$$

Equation (195-5) is the governing equation of **displacement v** and **deflection** θ of beams of physical constants M, I, and E. The first integration of (195-5) with respect to x, gives the deflection θ and displacement **v.**

For transverse loading **w** per unit length of the beam, the above differential equation becomes:

$$E.I \frac{d^4y}{dx^4} = w \tag{196-1}$$

Provided, **E** and **I** are constants.

A **concentrated load** W, at $x = a$, may be considered as a distributed load **w** per unit length of the beam such that

$$w = W\delta(x - a) = WU'(x - a) \tag{196-2}$$

Where, δ is the **Dirac delta function**.

The transform of (196-2) is

$$w(s) = We^{-as} \tag{196-3}$$

A **constant load** w_o per unit length in $0 < x < a$ and zero load for $x > a$ may be written
$$w = w_o[1 - U(x - a)] \tag{196-4}$$

The Laplace transform for constant load (196-4) is

$$w(s) = \frac{W}{s}\left(1 - e^{-as}\right)$$
(196-5)

The transform of a **couple of moment** μ applied at x = a is

$$\mu s e^{-as}$$
(196-6)

The inverse Laplace transform of (196-6)

$$\mu\delta'(x-a) = \mu U''(x-a)$$
(196-7)

Now consider the **general beam equation** (196-1)

$$E.I\frac{d^4y}{dx^4} = f(x)$$
(196-8)

where f(x) represents the **load per unit length** at a point x of the beam.

The subsidiary equation of (196-8) is given by

$$s^4 y(s) - s^3 y(0) - s^2 y'(0) - s y''(0) - y'''(0) = \frac{f(s)}{E.I}$$
(196-9)

i.e.,

$$y(s) = \frac{y(0)}{s} + \frac{y'(0)}{s^2} + \frac{y''(0)}{s^3} + \frac{y'''(0)}{s^4} + \frac{1}{E.I} f(s)$$
(196-10)

Inverting we get

$$y(x) = y(0) + y'(0)x + y''(0)\frac{x^2}{2!} + y''''(0)\frac{x^3}{3!} + \frac{1}{E.I} L^{-1}\{f(s)\}$$
(196-11)

In practical problems, some of the quantities y(0), y'(0), y"(0). y"'(0) are known if we are given at x = 0 the **deflection** or **slope** or **bending moment** or **shearing force**. The remaining quantities are determined from conditions at other points of the beam.

The following are applications of the above principles.

Example 127   **Deflection of beam under uniformly distributed load**

A beam is hinged at its ends x = 0 and x = l. It carries a **uniformly distributed load** w per unit length. Find the **static deflection** at any point on the beam.

Solution

244

$$E.I\frac{d^4y}{dx^4} = w_o \tag{197}$$

$$s^4y(s) - s^3y(0) - s^2y'(0) - sy''(0) - y'''(0) = \frac{w_o}{E.Is} \tag{197-1}$$

The initial conditions are

$$y(0) = y''(0) = 0$$

Put

$$y'(0) = A$$
$$y'''(0) = B$$

$$s^4y(s) = As^2 + B + \frac{w_o}{E.Is} \tag{197-2}$$

$$y(s) = \frac{A}{s^2} + \frac{B}{s^4} + \frac{w_o}{E.Is^5} \tag{197-3}$$

The inverse Laplace transform of (197-3) is

$$y(t) = L^{-1}\left\{\frac{A}{s^2} + \frac{B}{s^4} + \frac{w_o}{E.Is^5}\right\} = Ax + B\frac{x^3}{3!} + \frac{w_o}{E.I}\frac{x^4}{4!} \tag{197-4}$$

We have now to determine A and B from the other two conditions namely $y(l) = y''(l) = 0$,

$$0 = Al + B\frac{l^3}{3!} + \frac{w_o}{E.I}\frac{l^4}{4!} \tag{197-4.1}$$

$$0 = lB + \frac{w_o}{E.I}\frac{l^2}{2} \tag{197-4.2}$$

Therefore, the constants A and B are given by

$$B = -\frac{w_o}{E.I}\frac{l}{2} \tag{197-4.3}$$

$$A = \frac{w_o}{E.I}\frac{l^3}{24} \tag{197-4.4}$$

From (197-4.3) and (197-4.4), therefore, the **equation of displacement** (197-4) becomes

$$y(t) = \frac{w_o}{E.I}\frac{l^3}{24}x - \frac{w_o}{E.I}\frac{l}{12}x^3 + \frac{w_o}{E.I}\frac{1}{24}x^4$$

$$y(t) = \frac{w_o}{24E.I}x(l - x)(l^2 + lx - x^3) \tag{197-5}$$

Example 128   Deflection of beam under uniformly distributed load

A beam of length 2 l has both its ends built in. It carries a load w(x) per unit length such that

$$W(x) \quad = w_0 \qquad 0 < x < l$$
$$\quad = 0 \qquad l < x < 2l$$

Find the **static deflection** y(x) at any point x.

Solution

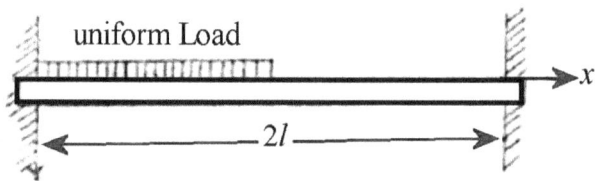

$$\frac{d^4 y}{dx^4} = \frac{w(x)}{E.I} = a.w(x) \tag{198}$$

$$s^4 y(s) - s^3 y(0) - s^2 y'(0) - sy''(0) - y'''(0) = \frac{aw_0}{s}\left(1 - e^{-k}\right) \tag{198-1}$$

Now y(0) = y'(0) = 0 and let y''(0) = A, y'''(0) = B

246

$$s^4 y(s) = As + B + \frac{aw_o}{s}\left(1 - e^{-ls}\right)$$ (198-2)

$$y(s) = \frac{A}{s^3} + \frac{B}{s^4} + \frac{aw_o}{s^5}\left(1 - e^{-ls}\right)$$ (198-3)

The inverse Laplace transform of (198-3) is

$$y(x) = A\frac{x^2}{2} + B\frac{x^3}{3!} + \frac{aw_o}{4!}\left[x^4 - U(x-l)(x-l)^4\right]$$ (198-4)

Where

$U(x-l) = 0$ where $\qquad 0 < x < l$

$U(x-l) = 1$ where $\qquad l < x < 2l$

To determine the constants A and B, we use the boundary conditions

$x = 2l$

$y = 0$

$y' = 0.$

From (198-4), put $x = 2l$ and $y = 0$, we get

$$0 = 2l^2.A + B\frac{4l^3}{3} + \frac{aw_o}{24}\left(16l^4 - l^4\right)$$

$$0 = 2.A + B\frac{4l}{3} + \frac{5aw_o l^2}{8}$$

$$A = -B\frac{2l}{3} - \frac{5aw_o l^2}{16}$$ (198-5)

The derivative of (198-4) gives

$$y'(x) = Ax + B\frac{x^2}{2} + \frac{aw_o}{4!}\left[4x^3 - 4(x-l)^3\right]$$ (198-6)

Put in (198-6), put $x = 2l$ and $y' = 0$, we get

$$0 = 2l.A + 2Bl^2 + \frac{aw_o}{24}\left(32l^3 - 4l^3\right)$$

$$A = -Bl - \frac{7aw_o l^2}{12}$$ (198-7)

Eliminate A between (198-5) and (198-7)

$$B = -\frac{13aw_o l}{16}$$ (198-7.1)

$$A = \frac{11.aw_o l^2}{48} \tag{198-7.2}$$

Hence, the equation of **static deflection** $y(x)$ at any point x, (198-4) becomes.

$$y(x) = \left(\frac{11.aw_o l^2}{96}\right)x^2 - \left(\frac{13aw_o l}{96}\right)x^3 + \frac{aw_o}{24}\left[x^4 - U(x - l)(x - l)^4\right] \tag{198-8}$$

Example 129   Deflection of beam under concentrated load

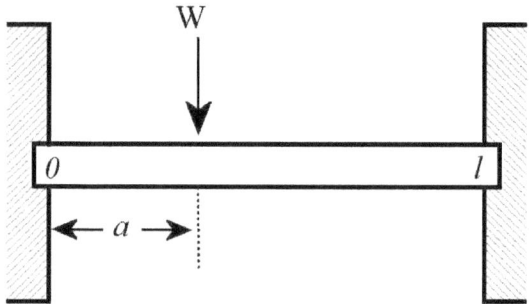

A beam of length l is clamped horizontally at its ends x = o and x = l and carries a concentrated load W at x = a. Find the **static deflection** y at any point x.

Solution

$$E.I\frac{d^4 y}{dx^4} = W\delta(x - a) \tag{199}$$

$$s^4 y(s) - s^3 y(0) - s^2 y'(0) - sy''(0) - y'''(0) = \frac{W}{E.Is}e^{-as} \tag{199-1}$$

The initial conditions

$$y(0) = y'(0) = 0$$
$$y''(0) = A$$
$$y'''(0) = B$$

$$s^4 y(s) = As + B + \frac{W}{EI}e^{-as} \tag{199-2}$$

$$y(x) = A\frac{x^2}{2} + B\frac{x^3}{3!} + \frac{W}{3!EI}U(x - a)(x - a)^3 \tag{199-3}$$

As before, we apply the initial conditions that determine the constants A and B

$y(0) = y(l) = 0$ since beam is fixed

$$A = \frac{Wa}{EI}\left(\frac{l-a}{l}\right)^2$$

$$B = -\frac{W}{EI}\frac{(l-a)^2(l+2a)}{l^3} \tag{199-4}$$

$y'(0) = y'(l) = 0$ since beam is fixed. The derivative of (199-3) gives

$$0 = Al + B\frac{l^2}{2} + \frac{W}{2EI}(l-a)^2$$

$$A = -B\frac{l}{2} - \frac{W}{2EI}\frac{(l-a)^2}{l} \tag{199-5}$$

The solution of (199-4) and (199-5) gives A and B as follows

$$A = -B\frac{l}{2} - \frac{W}{2EI}\frac{(l-a)^2}{l}$$

$$A = -B\frac{l}{2} - \frac{W}{2EI}\frac{(l-a)^2}{l}$$

Therefore, the deflection equation (199-3) becomes

$$y(x) = \frac{Wa}{2EI}\left(\frac{l-a}{l}\right)^2 x^2 - \frac{W}{6EI}\frac{(l-a)^2(l+2a)}{l^3}x^3 + \frac{W}{6EI}U(x-a)(x-a)^3 \quad (199\text{-}6$$

Example 130   Deflection of beam under terminal load

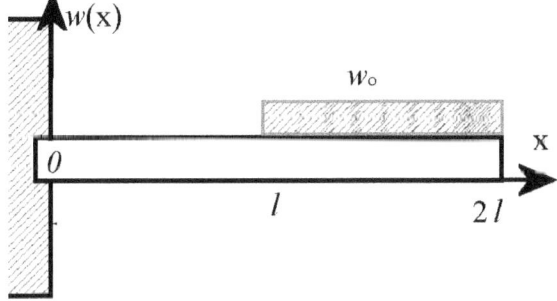

249

A cantilever of length, $2l$, has its end x=0 built in, while the end x=2 l is free. It carries a load w(x) per unit length which is zero over $0 < x < l$ and equal to a constant w_o over $l < x < 2l$. Find the static deflection at any point x of the cantilever.

$$E.I\frac{d^4y}{dx^4} = w(x) \tag{200}$$

Where $\quad$ w(x) $\quad = 0 \qquad\qquad 0 < x < l$
$\qquad\qquad\qquad\quad = w_o \qquad\qquad l < x < 2l$

Extend the definition of w(x) so as to be equal to w_o for all $x > o$. Therefore, equation (200) becomes

$$E.I\frac{d^4y}{dx^4} = \frac{1}{E.I}w(x) = aw(x) \tag{200-1}$$

$$s^4y(s) - s^3y(0) - s^2y'(0) - sy''(0) - y'''(0) = \frac{aw_o}{s}e^{-ls} \tag{200-2}$$

The initial conditions

$\qquad\qquad$ y(0) = y'(0) = 0
$\qquad\qquad$ y''(0) = A
$\qquad\qquad$ y'''(0) = B

Therefore,

$$s^4y(s) = As + B + \frac{aw_o}{s}e^{-ls} \tag{200-4}$$

$$y(x) = A\frac{x^2}{2} + B\frac{x^3}{3!} + \frac{aw_o}{24}U(x - l)(x - l)^4 \tag{200-5}$$

The constants A and B are determined *from* the conditions:

$\qquad\qquad$ Y''' (2l) = Y'' (2l) =0

since there is no bending shearing force at x = 2l

For x > l we have

$$y''(x) = A + Bx + \frac{aw_o}{2}(x - l)^2$$

$$y'''(x) = B + aw_o(x - l)$$

Therefore,

$$A = -2l.B - \frac{aw_o}{2}l^2 \tag{200-5.1}$$

$$B = -aw_ol \tag{200-5.2}$$

250

Or $\qquad A = \dfrac{3}{2} aw_o l^2$

Thus, the defelection equation (200-5) becomes

$$y(x) = \frac{3}{4} aw_o l^2 x^2 - \frac{1}{6} aw_o l.x^3 + \frac{aw_o}{24} U(x-l)(x-l)^4 \qquad (200\text{-}6)$$

Where

$$U(x-l) \qquad \begin{aligned} &= 0 && 0 < x < l \\ &= 1 && l < x < 2l \end{aligned}$$

Example 131   Deflection of beam under non-uniform load

A beam of length l is clamped horizontally at $x = o$ and freely hinged at $x = l$ It carries a load **wx** per unit length in $0 < x < l / 2$ and a load **w(l -x)** in $l / 2 < x < l$. Find the static deflection at any point x.

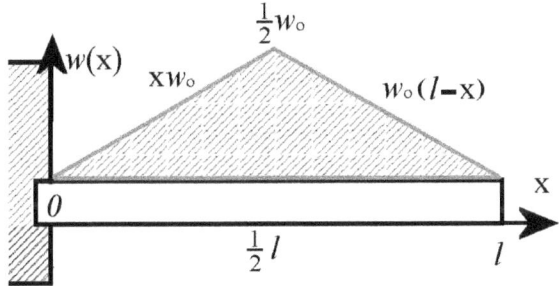

Solution

$$E.I \frac{d^4 y}{dx^4} = \left[wx - 2w\left(x - \frac{1}{2}\right) U\left(x - \frac{1}{2}\right)\right] \qquad (201)$$

$$s^4 y(s) - s^3 y(0) - s^2 y'(0) - s y''(0) - y'''(0) = \frac{w_o}{E.I}\left[\frac{1}{s^2} - \frac{2}{s^2} e^{-\frac{k}{2}}\right] \qquad (201\text{-}1)$$

The initial conditions

$$\begin{aligned} &y(0) = y'(0) = 0 \\ &y''(0) = A \\ &y'''(0) = B \end{aligned}$$

251

Therefore, equation (201-1) becomes

$$s^4 y(s) - As - B = \frac{w_o}{E.I.s^2}\left[1 - 2e^{-\frac{ls}{2}}\right]$$

(201-2)

Laplace Inversing (201-2), we get

$$y(x) = A\frac{x^2}{2} + B\frac{x^3}{6} + \frac{w_o x^5}{5!E.I} - \frac{2w_o}{5!E.I}\left(x - \frac{l}{2}\right)^5 U\left(x - \frac{l}{2}\right)$$

(201-3)

The constants A and B are determined as usual from the initial conditions, $y = y'' = 0$ when $x = l$.

First, $y(l) = 0$:

$$0 = A\frac{l^2}{2} + B\frac{l^3}{6} + \frac{w_o l^5}{5!E.I} - \frac{2w_o}{5!E.I}\left(l - \frac{l}{2}\right)^5$$

$$A = -B\frac{l}{3} - \frac{15}{8}\frac{w_o l^3}{5!E.I}$$

(201-3.1)

First, $y''(l) = 0$:

$$0 = A + Bl + \frac{w_o l^3}{3!E.I} - \frac{1}{8}\cdot\frac{2w_o l^3}{3!E.I}$$

$$A = -Bl - \frac{1}{8}\frac{w_o l^3}{E.I}$$

(201-3.2)

Eliminating and arranging, equations (201-3.1) and (201-3.2) give

$$B = -\frac{21}{128}\frac{w_o l^2}{E.I}$$

$$A = \frac{5}{128}\frac{w_o l^3}{E.I}$$

Substituting the values of A and B from these two equations in the expression for y(x), equation (201-3), we get

$$y(x) = \frac{5}{256}\frac{w_o l^3}{E.I}x^2 - \frac{7}{256}\frac{w_o l^2}{E.I}x^3 + \frac{w_o}{120E.I}x^5 - \frac{w_o}{60E.I}\left(x - \frac{l}{2}\right)^5 U\left(x - \frac{l}{2}\right)$$

(201-4)

Example 132   **Deflection of beam under non-uniform load and terminal force**

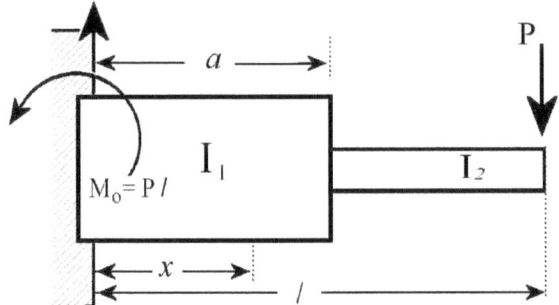

Determine the static deflection at any point of the cantilever beam shown above.

Solution

We shall use the **moment equation** instead of the **loading equation** and the change in the stiffness is accounted for by a function of the form.

$$\frac{1}{E.I}\left[1 + k.U(x - a)\right] \tag{202}$$

The **moment equation** for the element in the region $0 < x < a$, with moment of inertia I_1, is

$$\frac{d^2y}{dx^2} = \frac{M}{E.I_1}\left[1 - k.U(x - a)\right] \tag{202-1}$$

Where k is defined by

$$\frac{1}{E.I_1}(1 - k) = \frac{1}{E.I_2} \tag{202-1.1}$$

For which

$$k = 1 - \frac{I_1}{I_2}. \tag{202-1.2}$$

Therefore,

$$\frac{d^2y}{dx^2} = \frac{P(l - x)}{E.I_1}\left[1 - k.U(x - a)\right] \tag{202-3}$$

Since $y(0) = y''(0) = 0$ and

253

$$L\left\{ xU(x-a) \right\} = e^{-as}\left(\frac{a}{s} + \frac{1}{s^2} \right)$$ (202-3.1)

Therefore, the Laplace transform of (202-3), which is the subsidiary equation becomes

$$s^2 y(s) = \frac{P}{E.I_1}\left[\frac{l}{s} - \frac{1}{s^2} - \frac{l}{s}ke^{-as} + ke^{-as}\left(\frac{a}{s} + \frac{1}{s^2} \right) \right]$$ (202-4)

Laplace inverting (202-4), we get

$$y(x) = \frac{P}{E.I_1} L^{-1}\left[\frac{l}{s^3} - \frac{1}{s^4} - \frac{l}{s^3}ke^{-as} + ke^{-as}\left(\frac{a}{s^3} + \frac{1}{s^4} \right) \right]$$

$$= \frac{P}{E.I_1}\left(\frac{l.x^2}{2} - \frac{x^3}{6} - k\left(\frac{(l-a)(x-a)^2}{2} - \frac{(x-a)^3}{6} \right) \right).U(x-a)$$

Example 133   Critically loaded staut

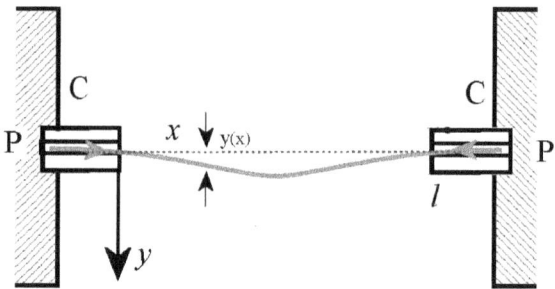

Find the critical loads for a strut clamped at both ends,

Solution

Let the fixing couple at each end be C. The **moment equation** becomes

$$E.I\ \frac{d^2 y}{dx^2} = C - Py$$ (203)

The Laplace transform of (203) is

$$E.I.[s^2 y(s) - sy(0) - y'(0)] = \frac{C}{s} - Py(s)$$

Therefore, $\quad [E.I.s^2 + P].y(s) = \dfrac{C}{s}$ $\hspace{4cm}$ (203-1)

$$y(s) = \frac{C}{s(E.I.s^2 + P)} = \frac{C}{E.I.s\left(s^2 + \dfrac{P}{E.I.}\right)} = \frac{C}{E.I.s(s^2 + n^2)}$$ $\hspace{1cm}$ (203-2)

Where

$$n^2 = \frac{P}{E.I.}$$ $\hspace{6cm}$ (203-3)

Therefore, the inverse Laplace transform of (203-2)

$$y(x) = L^{-1}\left\{\frac{C}{E.I.s(s^2 + n^2)}\right\} = \frac{C}{E.I}(1 - \cos nx)$$ $\hspace{1cm}$ (203-4)

From (203-3), equation (203-4) becomes

$$y(x) = \frac{C}{P}(1 - \cos nx)$$ $\hspace{5cm}$ (203-5)

For maximal loading condition, we need to find values of **n** that makes y(x) = 0.

At x = l, y = 0, we get

$$1 - \cos nl = 0$$ $\hspace{5.5cm}$ (203-6)

Therefore,

$$nl = 2\pi, 4\pi, \dots$$
$$n = 2\pi/l, 4\pi/l, \dots$$ $\hspace{5cm}$ (203-7)

First critical load occurs at n = n = $2\pi/l$, and is give by equation (203-3) by

$$P = n^2 E.I = \left(\frac{2\pi}{l}\right)^2 E.I$$ $\hspace{4cm}$ (203-8)

Example 134   **Non-uniform critically loaded columns**

The operational method for treating columns of several sections is by **shifting the origin** to the end of each section. Apply the **momentum** and **deflection equations** of beam subjected to loads to find the critical loads for the ith section of the non-uniform column shown in the figure

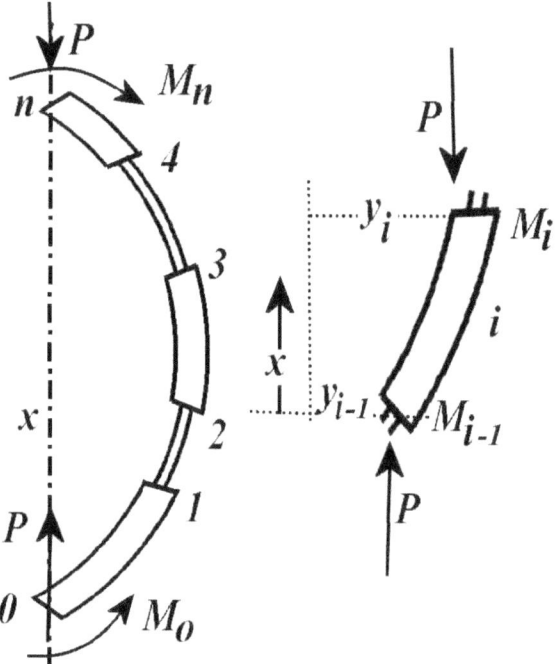

Solution

Consider the ith part of the column shown in the above diagram.

The **momentum equation** for the ith section gives

$$E_i.I_i \frac{d^2y}{dx^2} = -P[y(x) - y_{i-1}] + M_{i-1} \tag{204}$$

Dividing by $E_i.I_i$ and putting $P / E_i.I_i = a^2_i$, we get

$$\frac{d^2y}{dx^2} = -a_i^2(y(x) - y_{i-1}) + \frac{M_{i-1}}{E_i.I_i} \tag{204-1}$$

The Laplace transform of (204-1) is

$$s^2y(s) - sy_{i-1} - y'_{i-1} = -a_i^2\left(y(s) - \frac{y_{i-1}}{s}\right) + \frac{M_{i-1}}{E_i.I_i} \cdot \frac{1}{s} \tag{204-2}$$

Arranging the terms in (204-2), we gt

$$\left(s^2 + a_i^2\right)y(s) = sy_{i-1} + y'_{i-1} + a_i^2\frac{y_{i-1}}{s} + \frac{M_{i-1}}{E_i.I_i}.\frac{1}{s}$$

$$= \frac{s^2 + a_i^2}{s}y_{i-1} + y'_{i-1} + \frac{M_{i-1}}{E_i.I_i}.\frac{1}{s} \tag{204-3}$$

We have assumed that y_{i-1} is constant, such that we can determine the successive deflection.

$$y(s) = \frac{y_{i-1}}{s} + \frac{y'_{i-1}}{s^2 + a_i^2} + \frac{M_{i-1}}{E_i.I_i}.\frac{1}{s\left(s^2 + a_i^2\right)} \tag{204-4}$$

Inverting we get

$$y(x) = y_{i-1} + \frac{y'_{i-1}}{a_i}\sin a_i x + \frac{M_{i-1}}{E_i.I_i a_i^2}.(1 - \cos a_i x) \tag{204-5}$$

Now in this equation, and in the two equations obtained by differentiating it twice w.r.t x, put $x = l_i$ so as to obtain the **deflection**, **slope** and **bending moment** at **i** in terms of the corresponding quantities at **i -l**.

Deflection at x:: $\qquad y_i(x) = y_{i-1}\frac{y'_{i-1}}{a_i}\sin a_i l_i + \frac{M_{i-1}}{E_i.I_i a_i^2}.(1 - \cos a_i l_i) \tag{204-6}$

Slope at deflection: $\qquad y'_i(x) = y'_{i-1}\cos a_i l_i + \frac{M_{i-1}}{E_i.I_i a_i}.\sin a_i l_i \tag{204-7}$

Therefore, the bending moment is obtained dy differentiating the slope at deflection point x and multiplying by the moment of inertia I_i and the Young's modulus E_i, as follows

$$M_i(x) = E_i.I_i\frac{d^2 y_i(x)}{dx^2}$$

$$= E_i.I_i\frac{d}{dx}y'_i(x) = -y'_{i-1}E_i.I_i a_i \sin a_i l_i + M_{i-1}\cos a_i l_i \tag{204-8}$$

To sum up the problem, the deflection $y_i(x)$, slope $y'_i(x)$ and bending moment $M_i(x)$ may *be* expressed in **matrix form** as follows:

$$\begin{bmatrix} y_i \\ y'_i \\ M_i \end{bmatrix} = \begin{bmatrix} 1 & \dfrac{1}{a_i}\sin a_i l_i & \dfrac{1}{E_i.I_i a_i^2}(1 - \cos a_i l_i) \\ 0 & \cos a_i l_i & \dfrac{1}{E_i.I_i a_i}\sin a_i l_i \\ 0 & -E_i.I_i a_i \sin a_i l_i & \cos a_i l_i \end{bmatrix}\begin{bmatrix} y_{i-1} \\ y'_{i-1} \\ M_{i-1} \end{bmatrix} \tag{204-9}$$

In the same way, by repeated application, we can determine $y_{i-1}(x)$, slope $y'_{i-1}(x)$ and bending moment $M_{i-1}(x)$ in terms of $y_{i-2}(x)$, $y'_{i-2}(x)$ and $M_{i-2}(x)$, and so on.

Thus by multiplying the successive matrices we can obtain $y_n(x)$, $y'_n(x)$ and $M_n(x)$ in terms of $y_0(x)$, $y'_0(x)$ and $M_0(x)$ in the form

$$\begin{bmatrix} y_n \\ y'_n \\ M_n \end{bmatrix} = \begin{bmatrix} P_{11} & P_{12} & P_{13} \\ 0 & P_{22} & P_{23} \\ 0 & P_{32} & P_{33} \end{bmatrix} \begin{bmatrix} y_0 \\ y'_0 \\ M_0 \end{bmatrix} \qquad (204\text{-}10)$$

And, the **critical loads** can be obtained by substituting the **boundary conditions** in this equation.

Example 135   **Both ends of column built in**

Thus, for example in the case of a column whose ends are **both built in,** there is no **deflection** or **sloping**. Thus, we have

$$y_n(x) = y_0(x) = 0 \qquad (204\text{-}11)$$
$$y'_n(x) = y'_0(x) = 0 \qquad (204\text{-}12)$$

and since from (204-10), we have

$$y_n(x) = P_{11}\, y_0 + P_{12}\, y'_0 + P_{13}\, M_0 \qquad (204\text{-}13)$$

And $$y'_n(x) = P_{22}\, y'_0 + P_{23}\, M_0 \qquad (204\text{-}14)$$

Therefore, substituting from (204-11) and (204-12) into (204-13), we get

$$0 = P_{11}\, 0 + P_{12}\, 0 + P_{13}\, M_0 \qquad (204\text{-}13.1)$$

Therefore, substituting from (204-11) and (204-12) into (204-14), we get

$$0 = P_{22}\, 0 + P_{23}\, M_0 \qquad (204\text{-}13.2)$$

Therefore, from (204-13.1) and (204-13.2), the **matrix coefficients** for **built-in** ends are

$$P_{13} = 0$$
$$P_{23} = 0$$

Example 136   **Both ends of column hinged**

If the column is **hinged at both** ends, then **sloping is permitted**, but **deflection and momentum vanish**. Thus, we have

$$y_n(x) = y_o(x) = 0 \tag{204-14}$$
$$M_n = M_o = 0 \tag{204-15}$$

From equation (204-10), we have

$$y_n(x) = P_{11} y_o + P_{12} y'_o + P_{13} M_o \tag{204-16}$$
And
$$M_n = P_{32} y'_o + P_{33} M_o \tag{204-17}$$

Therefore, substituting from (204-14) and (204-15) in (204-16) and (204-17), we have the **matrix coefficients** for **hinged** ends

$$P_{12} = 0 \tag{204-18}$$
$$P_{32} = 0 \tag{204-19}$$

Example 137   Deflection of beam under elastic support

A uniform beam with its ends x= 0 and x =./ built in carries a **uniform load** w per unit length. At x = a, there is **an elastic support** which provides a reaction equal to λ times the deflection. Determine the **deflection** at any point x of the beam.

Solution

If R is the **reaction** at the support then

$$R = \lambda\, y(a) \tag{205}$$

$$E_i . I_i \frac{d^4 y}{dx^4} = -w - R\delta(x-a) \tag{205-1}$$

The Laplace transform of (205-1) is

$$E.I.\left[s^4 y(s) - s^3 y(0) - s^2 y'(0) - sy''(0) - y'''(0) \right] = \frac{w}{s} - Re^{-as} \tag{205-2}$$

The boundary conditions are

$$y(0) = y'(0) = 0$$
$$y''(0) = A$$
$$y'''(0) = B$$

Then

$$y(s) = \frac{A}{s^3} + \frac{B}{s^4} + \frac{w}{E.I.s^5} - \frac{R}{E.I.s^4} e^{-as} \tag{205-3}$$

Laplace Inversing (205-3), we get

$$y(t) = \frac{A}{2}x^2 + \frac{B}{6}x^3 + \frac{wx^4}{24E.I.} - \frac{R(x-a)^3}{6E.I.}U(x-a) \qquad (205\text{-}4)$$

The unknowns A, B, R can be determined as follows:

Since y(l) = y'(1)= 0, then form (205-4), we get

$$0 = \frac{A}{2}l^2 + \frac{B}{6}l^3 + \frac{wl^4}{24E.I.} - \frac{R(l-a)^3}{6E.I.} \qquad (205\text{-}4.1)$$

$$0 = \frac{A}{2}l^2 + \frac{B}{6}l^3 + \frac{wl^4}{24E.I.} - \frac{R(l-a)^3}{6E.I.} \qquad (205\text{-}4.2)$$

Since at x = a, y = R / λ then

$$\frac{R}{\lambda} = \frac{A}{2}a^2 + \frac{B}{6}a^3 + \frac{wa^4}{24E.I.} \qquad (205\text{-}4.3)$$

Equations (205-4.1), (205-4.2), and (205-4.3) determine the three constants A,B, and R, as we have doen repeatedly.

5.2. Exercises on Laplace Transform in practical applications

(1) A periodic e.m.f. **E sin ωt** is applied at time t = 0 to an RC circuit in series. If the initial current and charge are zero, find the current I at any instant t.

(2) In problem (1) replace **E sin ωt** by **E[U(t-t₀) –U(t-t₁)]**, $t_1 > t_0 > 0$.

(3) An e.m.f. E(t) is applied at time t o to an **RC** circuit in series. The initial current and charge are zero. Find the **charge and current** at any time t if

      (i) E is a constant, $E = E_0$
      (ii) $E = E_0 e^{-\alpha t}$
      (iii) $E = E_0 \delta(t)$, where $\delta(t)$ is the **Dirac delta** function.

(4) An electric circuit of an inductance L in series with a condenser of capacity C. At t = 0 an e.m.f. given by

$$E(t) \qquad = E_0 t\,/\,T_0 \qquad\qquad 0 < t < T_0$$
$$= 0 \qquad\qquad\qquad t > T_0$$

is applied. Assuming zero initial current and charge. Find the **charge** at any instant t.

(5) An e.m.f. **E cos (ωt + α)** is applied at time t = 0 to a series circuit of capacity C and inductance L. Find the current at any time t, assuming zero initial current and charge.

(6) In the circuit shown, calculate I_1, I_2, and I_3 if E is **constant** and the initial currents and charges are zero.

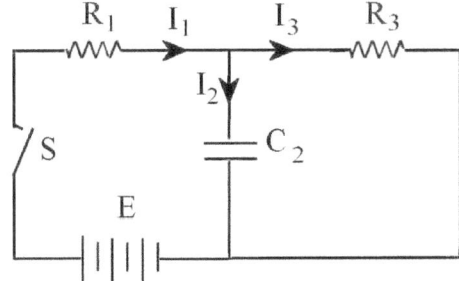

(7) In the circuit shown, the initial currents and charge are zero.
Find I_1 and I_2 when **V = E sin ωt**

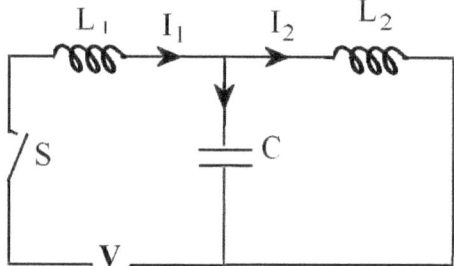

(8) In the circuit shown, set up the equations for the determination of the currents I_1, I_2, and I_3 and the charge Q_3 . Transform the problem into algebraic form, assuming E constant and initial currents and charges to be zero.

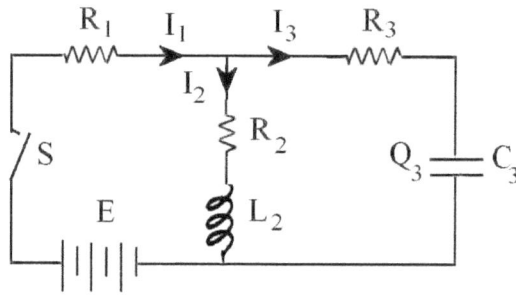

(9) A spring of stiffness k is placed on a smooth horizontal table. One end is fixed to a point O on the table and to the other end is attached a mass m. A force F(t), t > 0 acts on the mass. The differential equation of motion of the mass is

$$m \ddot{X} + kX = F(t)$$

(A) Assuming $X(0) = a$ and $\dot{X}(0) = 0$. Find X(t) under the following conditions

(i) $F(t) = F_0$ for t > 0, where F_0 is constant

(ii) $F(t) = F_0 e^{-\alpha t}$ where $\alpha > 0$

(iii) $F(t) = F_0 \sin \omega t$, where $\omega \neq \sqrt{\dfrac{k}{m}}$.

(iv) $F(t) = F_0 \sin \omega t$, where $\omega = \sqrt{\dfrac{k}{m}}$.

262

(v) $F(t) = F_0\,U(t-T)$

(vi) $F(t) = F_0\,\delta(t-T)$.

(B) Assuming $X(0) = \overset{..}{X}(0) = 0$. Find X(t) when

(vii) $F(t) = F_0\,\delta(t)$.

(10) A particle of **mass** m, at rest at the origin, is set in motion at t o by a blow of **impulse** P. Find its **displacement** at time t.

(11) A particle moves along a straight line such that its **displacement** x from a fixed point at time t is given by

$$\overset{...}{x} + 2\overset{..}{x} + 4x = 20\sin 4t$$

with initial conditions $x(0) = \overset{..}{x}(0) = 0$

Solve the equation and state which term in the result is the **transient term** and which is the steady state term. Find the **amplitude** and **period of the steady state**.

(12) Solve the equation of motion

$$m\overset{...}{x} = F(t)$$

of a particle with **initial conditions** $x(0) = \overset{..}{x}(0) = 0$, where F(t) is given by

F(t)	$= 2\,(F_0/T)\,t$	$0 < t < T/2$
	$= -2\,(F_0/T)\,-(t-T)$	$T/2 < t < T$
	$= 0$	$t > T$

(13) A beam with its ends **built in** at $x = 0$ and $x = l$ carries a **uniform load** w_0 per unit length. Find the **deflection** at any point x.

(14) Work the same problem in (13) with the end $x = 0$ **built in** and the end $x = l$ **hinged.**

(15) A beam with its ends $x = o$ and $x = l$ **hinged** carries a load w(x) per unit length given by:

w(x)	$= 0$	$0 < x < l/4$
	$= w_0$	$l/4 < x < l$

Find the **static deflection** at any point x.

(16) A cantilever beam with its end $x = 0$ **built in** and the end $x = l$ free carries a concentrated load P_0 at $x = 1/3$, find the **static deflection**.

(17) A beam with its ends $x = 0$ and $x = l$ **hinged** carries a concentrated load P_0 at $x = 1/4$. Find the deflection.

(18) A cantilever beam clamped at x = 0 and free at x = l carries a **uniform load** w_0 per unit length. Find the **deflection**.

CHAPTER 6

USING LAPLACE TRANSFORMATION IN SOLVING LINEAR PARTIAL DIFFERENTIAL EQUATIONS

The Laplace transformation can be used with advantage in solving **linear partial differential equations**. It is found that on applying this transformation the **partial differential equation** transforms to an ordinary differential equation. We illustrate by the following examples.

Example 138

Solve the partial differential equation

$$2x\frac{\partial Y(x,t)}{\partial t} + \frac{\partial Y(x,t)}{\partial x} = 2x \tag{206}$$

Given that

$$Y(x,0) = 1, \quad Y(0,t) = 1. \tag{206-1}$$

Solution

Writing equation (206) in the form

$$2x\, Y_t(x,t) + Y_x(x,t) = 2\,x \tag{206-2}$$

And noticing that

$$L\left\{ Y_x(x,t) \right\} = \int_0^\infty e^{-st} \frac{\partial Y(x,t)}{\partial x} dt$$

$$= \frac{\partial}{\partial x} \int_0^\infty e^{-st} Y(x,t)dt = y_x(x,s) \tag{206-3}$$

Equation (206-3) shows the Laplace transform of the derivative $Y_x(x,t)$, with respect to the variable t, yields the **derivative of the Lapalce transform** $y_x(x,s)$.

We get on Laplace transforming (206) with respect to t

$$2x[sy(x,s) - Y(x,0)] + y_x(x,s) = \frac{2x}{s} \tag{206-4}$$

With the initial condition that $Y(x,0) = 1$, we get

$$\frac{dy(x,s)}{dx} + 2xsy(x,s) = 2x + \frac{2x}{s} \tag{206-5}$$

This is a linear differential equation of the first order. The **integrating factor** (I.F.) is

$$\text{I.F.} = e^{\int 2xs\,dx} = e^{x^2s} \qquad\qquad (206\text{-}5.1)$$

Therefore, the solution of (206-5) becomes

$$y(x,s).\text{I.F.} = \int \text{I.F.}\left(2x + \frac{2x}{s}\right)dx + C$$

$$y(x,s)e^{sx^2} = \int 2x\left(1 + \frac{1}{s}\right)e^{sx^2}\,dx + C$$

$$= \frac{1}{s}\left(1 + \frac{1}{s}\right)e^{sx^2} + C$$

$$y(x,s) = \frac{1}{s}\left(1 + \frac{1}{s}\right) + Ce^{-sx^2} \qquad\qquad (206\text{-}6)$$

Laplace transforming (206-1) we get

$$y(0,s) = L\{\,Y(0,t)\} = 1/s \qquad\qquad (206\text{-}7)$$

From (206-6) and (206-7), the constant C is determined

$$y(0,s) = \frac{1}{s} = \frac{1}{s}\left(1 + \frac{1}{s}\right) + Ce^{-s0^2}$$

Therefore,

$$C = -\frac{1}{s^2} \qquad\qquad (206\text{-}8)$$

Thus, equation (206-6) becomes

$$y(x,s) = \frac{1}{s}\left(1 + \frac{1}{s}\right) - \frac{1}{s^2}e^{-sx^2} \qquad\qquad (206\text{-}9)$$

Inversing, we get

$$Y(x,t) = L^{-1}\left\{\frac{1}{s}\left(1 + \frac{1}{s}\right) - \frac{1}{s^2}e^{-sx^2}\right\} = 1 + t - (t - x^2)U(t - x^2) \qquad\qquad (206\text{-}10)$$

Where, $U(t - x^2) = 0$ when $\le t \le x^2$

$$Y(x,t) = 1 + t \qquad\qquad (206\text{-}10.1)$$

Where, $U(t - x^2) = 1$ when $t > x^2$

$$Y(x,t) = 1 + t - (t - x^2) = 1 + x^2 \qquad\qquad (206\text{-}10.2)$$

Example 139

Find the solution of

$$\frac{\partial U(x,t)}{\partial x} = 2\frac{\partial U(x,t)}{\partial t} + U(x,t) \tag{207}$$

$$U(x,0) = 4e^{-2x} \tag{207-1}$$

which is bounded for $x > 0, t > 0$.

Solution

Writing equation (207) in the form

$$U_x(x,t) = 2U_t(x,t) + U(x,t)$$

Therefore, the Laplace transform of (207) is

$$u_x(x,s) = 2[su_t(x,s) - U(x,0)] + u(x,s) \tag{207-2}$$

From (207-1) and (207-2), we get

$$u_x(x,s) = 2[su(x,s) - 4e^{-2x})] + u(x,s)$$

$$u_x(x,s) = u(x,s)(2s+1) - 8e^{-2x}$$

$$\frac{du(x,s)}{dx} = u(x,s)(2s+1) - 8e^{-2x} \tag{207-3}$$

which is a **linear differential equation** of the first order.

Integrating factor is

$$\text{I.F.} = e^{\int -(2s+1)dx} = e^{-(2s+1)x} \tag{207-4}$$

$$u(x,s)e^{-(2s+1)x} = -8\int e^{-(2s+1)x}e^{-2x}dx + C$$

$$= -8\int e^{-(2s+3)x}dx + C = \frac{8}{2s+3}e^{-(2s+3)x} + C \tag{207-5}$$

Therefore, from (207-5), we get

$$u(x,s) = \frac{8e^{-2x}}{2s+3} + Ce^{(2s+1)x} \tag{207-6}$$

Since $U(x,t)$ must be bounded as $x \rightarrow \infty$, we must have $u(x,s)$ also bounded as $x \rightarrow \infty$,

Therefore,
$$C = 0 \qquad\qquad (207\text{-}6.1)$$

Laplace inversing (207-6), we get

$$U(x,t) = L^{-1}\left\{\frac{8e^{-2x}}{2s+3}\right\} = 4e^{-2x}L^{-1}\left\{\frac{1}{s+\dfrac{3}{2}}\right\} = 4e^{-2x-\frac{3}{2}t} \qquad\qquad (207\text{-}7)$$

Example 140

Solve the partial differential equation

$$x\frac{\partial Y(x,t)}{\partial x} + \frac{\partial Y(x,t)}{\partial t} + U(x,t) = xF(t) \qquad\qquad (208)$$

with the boundary conditions

$$Y(x,0) = Y(0,t) = 0 \qquad\qquad (208\text{-}1)$$

Solution

The Laplace transforming (208) with respect to t is

$$xy_x(x,s) + sy(x,s) - Y(x,0) + y_x(x,s) = xf(s) \qquad\qquad (208\text{-}2)$$

Therefore, with $Y(x,0) = 0$,

$$x\frac{dy(x,s)}{dx} + sy(x,s) + y_x(x,s) = xf(s)$$

i.e.,

$$\frac{dy(x,s)}{dx} + \frac{s+1}{x}y(x,s) = f(s) \qquad\qquad (208\text{-}3)$$

The Integrating Factor is

$$\text{I.F.} = e^{\int \frac{s+1}{x}dx} = e^{(s+1)\log x} = x^{s+1} \qquad\qquad (208\text{-}4)$$

Therefore, the solution (208-3) is

$$y(x,s)x^{s+1} = \int f(s)x^{s+1}dx + C \qquad (208\text{-}5)$$

$$= f(s)\frac{x^{s+2}}{s+2} + C$$

Therefore,

$$y(x,s) = f(s)\frac{x}{s+2} + \frac{C}{x^{s+1}} \qquad (208\text{-}6)$$

Since $Y(0,t) = 0$, therefore, $y(0,s) = 0$,

$$C = 0 \qquad (208\text{-}6.1)$$

The inverse Laplace transform of (208-6) is

$$Y(x,t) = L^{-1}\left\{f(s)\frac{x}{s+2}\right\} \qquad (208\text{-}7)$$

Where the convolution integral takes the form

$$Y(x,t) = x\left\{e^{-2t} * F(t)\right\}$$

$$= x\int_0^t e^{-2(t-\lambda)}F(\lambda)d\lambda$$

$$= xe^{-2t}\int_0^t e^{2\lambda}F(\lambda)d\lambda \qquad (208\text{-}8)$$

Example 141

Solve the partial differential equation

$$U_{xx}(x,t) - 2U_{tx}(x,t) + U_{tt}(x,t) = 0 \qquad (0 < x < 1 , t > 0) \qquad (209)$$

with the boundary conditions

$$U(x,0) = U_t(x,0) = U(0,t) = 0$$
$$U(1,t) = F(t) \qquad\qquad t > 0 \qquad (209\text{-}1)$$

Solution

The Laplace transform of the second derivative $U_{xx}(x,t)$ is

$$L\left\{U_{xx}(x,t)\right\} = \int_0^\infty e^{-st}\frac{\partial^2}{\partial x^2}U(x,t)dt \qquad (209\text{-}2)$$

269

$$= \frac{\partial^2}{\partial x^2} \int_0^\infty e^{-st} U(x,t)dt = \frac{\partial^2}{\partial x^2} u(x,s) = u_{xx}(x,s) \qquad (209\text{-}3)$$

Therefore, from the boundary conditions and (209-3), equation (209) is Laplace transformed to

$$u_{xx}(x,s) - 2\frac{\partial}{\partial x}[su(x,s) - U(x,0)] + s^2u(x,s) - sU(x,0) - U_t(x,0) = 0$$

$$u_{xx}(x,s) - 2su_x(x,s) + s^2u(x,s) = 0 \qquad (209\text{-}4)$$

We could also use the Algebraic form of derivatives to write (209-4) as follows

$$(D^2 - 2\,s\,D + s^2)\,u = 0$$

$$(D\text{-}s)^2\,u = 0 \qquad (209\text{-}5)$$

Where is the differential operator $D = d/dx$.

Therefore,

$$u(x,s) = e^{sx}(A + Bx) \qquad (209\text{-}6)$$

From the initial conditions, we have:

First, since $U(0,t) = 0$, therefore, $u(0,s) = 0$ and hence $A = 0$.

Therefore, $\qquad u(x,s) = Bxe^{sx} \qquad (209\text{-}6.1)$

Second, since $U(1,t) = F(t)$, therefore $u(1,s) = f(s)$

$$f(s) = Be^s$$

i.e., $\qquad B = f(s)e^{-s} \qquad (209\text{-}6.2)$

Thus, equations, equation (209-6) becomes

$$u(x,s) = xf(s)e^{-(1-x)s} \qquad (209\text{-}6.3)$$

The inverse Laplace transform of (209-6.3) is

$$U(x,t) = L^{-1}\left\{xf(s)e^{-(1-x)s}\right\} = xF(t + x - 1)U(t + x - 1) \qquad (209\text{-}7)$$

Where $U(t + x - 1)$ is the unit step function, while the bi-variate $U(x,t)$ the sought function. Thus,

$U(x,t)$	$= 0$	$x < t < 1\text{-}x$
	$= x\,F(t + x\text{-}1)$	$t > 1\text{-}x$

Example 142

Find a bounded solution of

$$\frac{\partial V(x,t)}{\partial t} = k\frac{\partial^2 V(x,t)}{\partial x^2} \qquad x > 0,\, t > 0 \qquad\qquad (210)$$

subject to the boundary conditions $V(0,t) = F(t)$, $V(x,0) = 0$

Solution

The Laplace transform of (210) is

$$sv(x,s) - V(x,0) = kv_{xx}(x,s) \qquad\qquad (210\text{-}1)$$

This can also be written as

$$\frac{d^2 v(x,s)}{dx^2} - \frac{s}{k}v(x,s) = 0 \qquad\qquad (210\text{-}2)$$

The solution of (210-2) is

$$v(x,s) = Ae^{\sqrt{\frac{s}{k}}x} + Be^{-\sqrt{\frac{s}{k}}x} \qquad\qquad (210\text{-}3)$$

Since the solution is bounded, then A = 0, therefore,

$$v(x,s) = f(s)e^{-\sqrt{\frac{s}{k}}x} \qquad\qquad (210\text{-}3.1)$$

Before Laplace inversing (210-3.1), we consider the inversion of the exponential part

$$L^{-1}\left\{e^{-\sqrt{\frac{s}{k}}x}\right\} = \frac{1}{2\sqrt{\pi kt^3}}e^{-\frac{x^2}{4t}} \qquad\qquad (210\text{-}4.1)$$

Then, we consider the convolution integral, therefore

$$V(x,t) = F(t) * \frac{x}{2\sqrt{\pi kt^3}}e^{-\frac{x^2}{4t}} \qquad\qquad (210\text{-}4.2)$$

$$V(x,t) = \frac{x}{2\sqrt{\pi kt^3}}\int_0^t \frac{F(\lambda)}{(t-\lambda)^{\frac{3}{2}}}e^{-\frac{x^2}{4(t-\lambda)}}d\lambda \qquad\qquad (210\text{-}5)$$

6.1. Transverse vibrations of a stretched string under gravity.

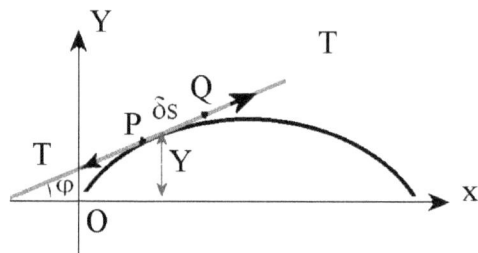

Let **T** be the tension in the string and **m** mass per unit length.
Consider an element PQ of the string of length δs.

The component of the tension at P resolved in the direction OY is

$$- T \sin \varphi = -T \frac{\partial Y}{\partial s} \tag{211-1}$$

The component of the tension at Q resolved in the direction OY is

$$T \frac{\partial Y}{\partial s} + \frac{\partial}{\partial s} \left(T \frac{\partial Y}{\partial s} \right) \delta s \tag{211-2}$$

Resultant force due to the two tensions in the direction OY is

$$\frac{\partial}{\partial s} \left(T \frac{\partial Y}{\partial s} \right) \delta s = T \frac{\partial^2 Y}{\partial s^2} \delta s \tag{211-3}$$

Therefore, the balance of external and internal forces with induced acceleration forces gives

$$m\delta s \frac{\partial^2 Y}{\partial t^2} = T \frac{\partial^2 Y}{\partial s^2} \delta s - m\delta s g \tag{211-4}$$

Put

$$a^2 = T / m \tag{211-4.1}$$

$$\frac{\partial^2 Y}{\partial t^2} = a^2 \frac{\partial^2 Y}{\partial s^2} - g \tag{211-5}$$

Assuming small lateral displacements and gradients, we can replace derivative of s by derivative of x as follows

$$\frac{\partial^2 Y}{\partial s^2} = \frac{\partial^2 Y}{\partial x^2}$$ (211-5.1)

Therefore, (211-5) becomes

$$\frac{\partial^2 Y}{\partial t^2} = a^2 \frac{\partial^2 Y}{\partial x^2} - g$$ (211-5.2)

i.e., $\qquad Y_{tt}(x,t) = a^2 Y_{xxt}(x,t) - g$ (211-5.3)

Example 143

A semi - infinite stretched string of negligible weight has its distant end fixed while the end $x = 0$ is initially at the origin and moves along the Y-axis such that

$$Y(t) - C \sin \omega t, \ t > 0.$$

The string is initially along the x-axis with no initial velocity.
When the end $x = 0$ starts to move find the **shape of the string** at any subsequent instant.

Solution

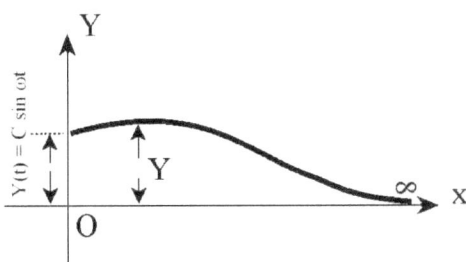

Differential equation of motion of the siring is

$$Y_{tt}(x,t) = a^2 Y_{xxt}(x,t)$$ (212)

The initial conditions, at $t = 0$ displacements and velocities are zero.

$$Y(x,0) = Y_t(x,0) = 0$$ (212-1)

The boundary, at $x = 0$ is given by

$$Y(t) = C \sin \omega t, \ t > 0 \qquad\qquad (212\text{-}2)$$

By definition, at $x = \infty$

$$\lim_{x \to \infty} Y(x,t) = 0 \qquad\qquad (212\text{-}3)$$

Transforming (212) with respect to t we get

$$s^2 y(x,s) - sY(x,0) - Y_t(x,0) = a^2 y_{xx}(x,s) \qquad\qquad (212\text{-}4)$$

Since, from (212-1), $Y(x,0) = Y_t(x,0) = 0$, therefore, (212-5) becomes

$$s^2 y(x,s) = a^2 y_{xx}(x,s) \qquad\qquad (212\text{-}5)$$

Therefore, Laplace transform of the temporal solution of $Y(x,t)$ is

$$y(x,s) = Ae^{\frac{s}{a}x} + Be^{-\frac{s}{a}x} \qquad\qquad (212\text{-}6)$$

Equation (212-3) transforms into

$$L\left\{ \lim_{x \to \infty} Y(x,t) \right\} = y(\infty,s) = 0 \qquad\qquad (212\text{-}7)$$

From (212-6) and (212-7), the constant

$$A = 0 \qquad\qquad (212\text{-}7.1)$$

Similarly, equation (212-2) transforms into

$$y(0,s) = L\left\{ C \sin \omega t \right\} = \frac{C\omega}{s^2 + \omega^2} \qquad\qquad (212\text{-}7.2)$$

Thus, from equations (212-6), (212-7.1) and (212-7.2) , we get

$$B = \frac{C\omega}{s^2 + \omega^2} \qquad\qquad (212\text{-}7.3)$$

And, equation (212-6) becomes

$$y(x,s) = \frac{C\omega}{s^2 + \omega^2} e^{-\frac{s}{a}x} \qquad\qquad (212\text{-}8)$$

The inverse Laplace transform of (212-8) is

$$Y(x,t) = L^{-1}\left\{\frac{C\omega}{s^2 + \omega^2}e^{-\frac{s}{a}x}\right\} = C\sin\left(\omega\left(t - \frac{x}{a}\right)\right).U\left(t - \frac{x}{a}\right) \qquad (212\text{-}9)$$

Where,

$$U\left(t - \frac{x}{a}\right) \quad = 0 \qquad\qquad < t < x/a$$

$$= 1 \qquad\qquad t > x/a \qquad\qquad\qquad (212\text{-}9.1)$$

The motion may be interpreted as follows:

A point at distance x from the origin remains at rest until a time $t = x/a$ elapses, when the motion at the origin is transferred to the point. The time $t = x/a$ is the time necessary for a **wave traveling** with velocity a to describe a distance x.

Example 144

A semi - infinite string has its end $x = 0$ fixed while the distant and is looped around a vertical support that cannot exert any vertical force upon the string i.e.,

$$T\frac{\partial Y}{\partial x} = 0 \qquad\qquad\qquad\qquad (213)$$

at that end. The string is initially supported along the x – axis and at time $t = 0$ the support is removed and the string moves downwards under the action of gravity.
Find the **shape of the string** at any instant.

Solution

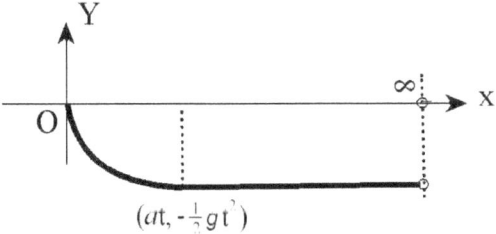

$$Y_{tt}(x,t) = a^2 Y_{xxt}(x,t) - g \qquad\qquad\qquad (214)$$

At time $t = 0$, displacements and velocities at all points of the string are zero.

Therefore, $Y(x,0) = Y_t(x,0) = 0$ (214-1)

At the end $x = 0$, $Y = 0$ for all t, i.e.,

$$Y(0,t) = 0$$ (214-2)

At the distant end $x = \infty$, the vertical component of the tension is zero.

$$\lim_{x \to \infty} Y_x(x,t) = 0$$ (214-3)

Transforming (214) with respect to t

$$s^2 y(x,s) = a^2 y_{xx}(x,s) - \frac{g}{s}$$ (214-4)

$$\frac{d^2 y(x,s)}{dx^2} - \frac{s^2}{a^2} y(x,s) = \frac{g}{a^2 s}$$ (214-4.1)

Which solution is

$$y(x,s) = Ae^{\frac{s}{a}x} + Be^{-\frac{s}{a}x} - \frac{g}{s^3}$$ (214-5)

Equation (214-3) transforms into

$$y(\infty,s) = L\left\{ \lim_{x \to \infty} Y_x(x,t) \right\} = 0$$

Which, upon substitution in (214-5), gives

$$A = 0$$ (214-5.1)

Similarly, equation (214-2) transforms into

$$y(0,s) = L\left\{ Y(0,t) \right\} = 0$$ (214-5.2)

Upon substitution by (214-5.1) and (214-5.2) in (214-5), gives

$$B = \frac{g}{s^3}$$ (214-5.3)

Therefore, from (214-5.1) and (214-5.3), equation (214-5) becomes

$$y(x,s) = \frac{g}{s^3}\left(e^{-\frac{s}{a}x} - 1 \right)$$ (214-6)

The inverse Laplace transform of (214-6) is

$$Y(x,t) = L^{-1}\left\{\frac{g}{s^3}\left(e^{-\frac{s}{a}x}-1\right)\right\} = gL^{-1}\left\{\frac{1}{s^3}e^{-\frac{s}{a}x}\right\} - gL^{-1}\left\{\frac{1}{s^3}\right\}$$

$$= \frac{1}{2}g\left(t-\frac{x}{a}\right)^2 . U\left(t-\frac{x}{a}\right) - g\frac{t^2}{2}$$ (214-7)

Where,

(i) $U\left(t-\frac{x}{a}\right)$ $= 0$ $0 < t < x/a$

$$Y(x,t) = -g\frac{t^2}{2}$$ (214-7.1)

(ii) $U\left(t-\frac{x}{a}\right)$ $= 1$ $t > x/a$

$$Y(x,t) = \frac{1}{2}g\left(t-\frac{x}{a}\right)^2 - g\frac{t^2}{2}$$

$$Y(x,t) = \frac{1}{2}g\left(t-\frac{x}{a}\right)^2 = \frac{g}{2}\left(\frac{x}{a}\right)^2 - \frac{1}{2}gt^2$$ (214-7.2)

Considering equation (214-7.2)

$$x^2 - 2ax = \frac{2a^2 Y(x,t)}{g}$$

$$(x-at)^2 = \frac{2a^2}{g}\left(Y(x,t)+\frac{1}{2}gt^2\right)$$ (214-8)

Which is a parabola vertex (at, - gt^2 / 2), and whose axis is parallel to OY.

Considering equation (214-7.1), where $Y(x,t) = -g\frac{t^2}{2}$, in the instantaneous position of the string, we notice that, at any time t, all elements of the string to the right of the point **x > at,** move like **freely falling bodies,** because the acceleration g is the only driving force in generating the displacement Y(x,t).

Example 145

A semi infinite stretched string has its distant end fixed on the x - axis while the end x = 0 is looped around the Y - axis which exerts, no vertical force on the loop. The string is initially at rest in the position $Y = e^{-x}$ released from this position with no external forces acting on it. Find the displacement of the string at any point x and at any instant t.

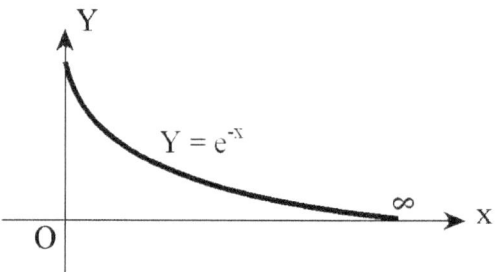

Solution

The equation of motion of the string is

$$Y_{tt}(x,t) = a^2 Y_{xxt}(x,t) \qquad (215)$$

Boundary conditions are

$$Y(x,0) = e^{-x} \qquad (215\text{-}1.1)$$
$$Y_t(x,0) = 0 \qquad (215\text{-}1.2)$$
$$Y_x(0,t) = 0 \qquad (215\text{-}1.3)$$
$$\lim_{x \to \infty} Y_x(x,t) = 0 \qquad (215\text{-}1.4)$$

Transforming (215) with respect to t

$$s^2 y(x,s) - sY(x,0) - Y_t(x,0) = a^2 y_{xx}(x,s) \qquad (215\text{-}2)$$

From (215-1.1) and (215-2), we get

$$s^2 y(x,s) - se^{-x} = a^2 y_{xx}(x,s) \qquad (215\text{-}2.1)$$

i.e.,
$$\frac{d^2 y(x,s)}{dx^2} - \frac{s^2}{a^2} y(x,s) = -\frac{s}{a^2} e^{-x} \qquad (215\text{-}2.2)$$

With the solution

$$y(x,s) = Ae^{\frac{s}{a}x} + Be^{-\frac{s}{a}x} + \frac{1}{D^2 - \frac{s^2}{a^2}}\left(-\frac{s}{a^2}e^{-x}\right)$$

$$= Ae^{\frac{s}{a}x} + Be^{-\frac{s}{a}x} - \frac{s}{a^2 - s^2}e^{-x} \tag{215-3}$$

Since, $D^2(e^x) = 1$.

As usual, the two constants A and B in equation (215-3), are determined from the initial and boundary conditions in (215-1). Thus,

$$y(\infty,s) = L\left\{\lim_{x\to\infty} Y_x(x,t)\right\} = 0$$

Therefore, $A = 0$ $\hspace{4cm}$ (215-3.1)

Thus, (215-3) becomes

$$y(x,s) = Be^{-\frac{s}{a}x} - \frac{s}{a^2 - s^2}e^{-x} \tag{215-3.2}$$

And from (215-1.3), $Y_x(0,t) = 0$

$$y_x(0,s) = L\{Y_x(0,t)\} = 0$$

Therefore, differentiating (215-3.2) and substituting by $y_x(0,s) = 0$, we get

$$y_x(0,s) = -\frac{s}{a}Be^{-\frac{s}{a}x} + \frac{s}{a^2 - s^2}e^{-x} = -\frac{s}{a}B + \frac{s}{a^2 - s^2} = 0$$

Thus,

$$B = \frac{a}{a^2 - s^2} \tag{215-3.3}$$

Therefore, from (215-3.2) and (215-3.3), equation (215-3) becomes

$$y(x,s) = -\frac{a}{a^2 - s^2}e^{-\frac{s}{a}x} + \frac{s}{a^2 - s^2}e^{-x} \tag{215-3.4}$$

The inverse Laplace transform of (215-3.4) is

$$Y(x,t) = L^{-1}\left\{-\frac{a}{a^2 - s^2}e^{-\frac{s}{a}x} + \frac{s}{a^2 - s^2}e^{-x}\right\}$$

$$= -U(at - x)\sinh(at - x) + e^{-x}\cosh at \tag{215-4}$$

Where,

$$U(at - x) \quad = 0 \qquad t < x/a$$
$$= 1 \qquad t > x/a$$

Example 146

A string is stretched between the two fixed points $x = 0$ and $x = 1$. The string has initially the shape

$$Y = b \sin \frac{\pi x}{1}$$

And is released from rest at $t = 0$ in that position. Find **the shape** of the string at any subsequent instant.

Solution

The differential equation of motion of the string is

$$Y_{tt}(x,t) = a^2 Y_{xxt}(x,t) \qquad 0 < x < l, \qquad t > 0 \qquad\qquad (216)$$

The boundary conditions are

$$Y(x,0) = b \sin \pi x / l \qquad\qquad (216\text{-}1.1)$$
$$Y_t(x,0) = 0 \qquad\qquad (216\text{-}1.2)$$
$$Y(0,t) = 0 \qquad\qquad (216\text{-}1.3)$$
$$Y(l,t) = 0 \qquad\qquad (216\text{-}1.4)$$

Transforming (216) with respect to t

$$s^2 y(x,s) - sY(x,0) - Y_t(x,0) = a^2 y_{xx}(x,s) \qquad\qquad (216\text{-}2)$$

From (216.1) and 216-2), we get

$$s^2 y(x,s) - s.b \sin\left(\frac{x\pi}{1}\right) = a^2 y_{xx}(x,s) \qquad\qquad (216\text{-}2.1)$$

i.e.,

$$\frac{d^2 y}{dx^2} - \frac{s^2}{a^2} y(x,s) = -\frac{sb}{a^2} \sin\left(\frac{x\pi}{1}\right) \qquad\qquad (216\text{-}2.2)$$

With the solution

$$y(x,s) = A \cosh\frac{s}{a}x + B\sinh\frac{s}{a}x + \frac{1}{D^2 - \frac{s^2}{a^2}}\left(-\frac{bs}{a^2}\sin\frac{\pi}{1}x\right)$$

280

$$= A \cosh \frac{s}{a} x + B \sinh \frac{s}{a} x + \frac{1}{-\frac{\pi^2}{1^2} - \frac{s^2}{a^2}} \left(-\frac{bs}{a^2} \sin \frac{\pi}{1} x \right)$$

$$= A \cosh \frac{s}{a} x + B \sinh \frac{s}{a} x + \frac{bl^2 s}{a^2 \pi^2 + l^2 s^2} \sin \frac{\pi}{1} x \qquad (216\text{-}3)$$

From the boundary condition (216-1.3), Y(0,t) = 0 , therefore, y(0,s) = 0 and A = 0.

Thus

$$y(x,s) = B \sinh \frac{s}{a} x + \frac{bl^2 s}{a^2 \pi^2 + l^2 s^2} \sin \frac{\pi}{1} x \qquad (216\text{-}3.1)$$

From the boundary condition (216-1.4), Y(l,t) = 0 , therefore, y(l,s) = 0

Thus

$$y(l, s) = B \sinh \frac{s}{a} l + \frac{bl^2 s}{a^2 \pi^2 + l^2 s^2} \sin \frac{\pi}{1} l = 0$$

Thus, $B = 0$

And

$$y(x,s) = \frac{bl^2 s}{a^2 \pi^2 + l^2 s^2} \sin \frac{\pi}{1} x$$

$$= \frac{bs}{\frac{a^2 \pi^2}{1^2} + s^2} . \sin \frac{\pi}{1} x \qquad (216\text{-}3.2)$$

The inverse Laplace transform of (216-3.2) is

$$Y(x,t) = L^{-1} \left\{ \frac{bs}{\frac{a^2 \pi^2}{1^2} + s^2} . \sin \frac{\pi}{1} x \right\} = b \sin \frac{\pi}{1} x \cos \frac{\pi a}{1} t \qquad (216\text{-}4)$$

6.2. Longitudinal vibrations of bars

Let one end O of an elastic bar be fixed and let A be the cross sectional area of the bar, ρ its density and **E** its Young's modulus.

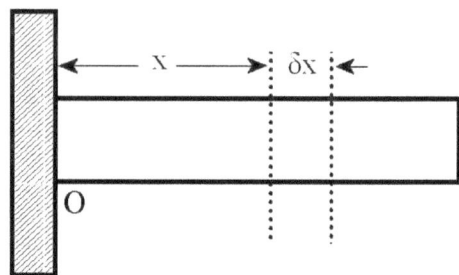

Consider an element of length **δx** of the bar, Its mass is **ρ A δx**.

Suppose that the length x extends a distance Y then a length δx extends a distance

$$\frac{\partial Y}{\partial x} \delta x \qquad (217)$$

The Strain, Stress, and Tension in the element are

Strain at x $\qquad \left(\frac{\partial Y}{\partial x} \delta x\right) / \delta x = \frac{\partial Y}{\partial x} \qquad$ (217-1)

Stress at x $\qquad E\frac{\partial Y}{\partial x} \qquad$ (217-2)

Tension at x $\qquad EA\frac{\partial Y}{\partial x} \qquad$ (217-3)

At distance x + δx, the Tension is

$$EA\frac{\partial Y}{\partial x} + \frac{\partial}{\partial x}\left(EA\frac{\partial Y}{\partial x}\right)\delta x \qquad (217-4)$$

Therefore, **resultant tension** in the element δx

$$EA\frac{\partial^2 Y}{\partial x^2} \delta x \qquad (217-5)$$

The **induced inertial force** is

$$\rho A\frac{\partial^2 Y}{\partial t^2} \delta x \qquad (217-6)$$

Therefore,

$$\rho A \frac{\partial^2 Y}{\partial t^2} \delta x = EA \frac{\partial^2 Y}{\partial x^2} \delta x$$

$$\frac{\partial^2 Y}{\partial t^2} = \frac{E}{\rho} \frac{\partial^2 Y}{\partial x^2} \qquad\qquad (217\text{-}7)$$

Put $\qquad\qquad a^2 - E / \rho \qquad\qquad\qquad\qquad\qquad (217\text{-}7.1)$

$$Y_{tt}(x,t) = a^2 Y_{xx}(x,t) \qquad\qquad\qquad (217\text{-}7.2)$$

Example 147

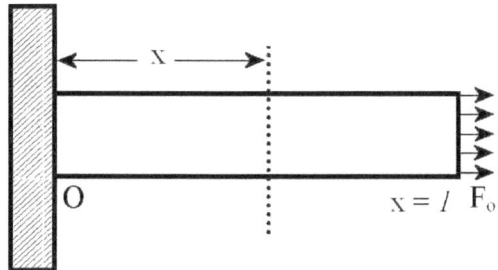

$x = l \quad F_o$

An elastic bar of length l, has its end $x = 0$ fixed while the other end $x = l$ is acted upon by a constant force F_o per unit area parallel to the bar. The bar is initially unstrained and at rest. Find the **longitudinal displacement** of the end $x = l$ at any instant.

Solution

Equation (217-7.2) is subjected to the following initial and boundary conditions:

At t = 0 $\qquad$ $Y(x,0)$ $\quad = 0$ $\qquad\qquad\qquad\qquad\qquad$ (218)
$\qquad\qquad\quad$ $Y_t(x,0)$ $\quad = 0$ $\qquad\qquad\qquad\qquad\qquad$ (218-1)
At x = 0 $\qquad$ $Y(0,t)$ $\quad = 0$ $\qquad\qquad\qquad\qquad\qquad$ (218-2)
At x = l $\qquad$ Stress F_o $= EY_x(l,t)$ $\qquad\qquad\qquad$ (218-3)

Transforming (217-7.2), we get

$$s^2 y(x,s) - sY(x,0) - Y_t(x,0) = a^2 y_{xx}(x,s) \qquad\qquad (218\text{-}4)$$

From the initial conditions given above, (218-4) becomes

$$s^2 y(x,s) = a^2 y_{xx}(x,s)$$

$$\frac{d^2 y(x,s)}{dx^2} = \frac{s^2}{a^2} y_{xx}(x,s) \tag{218-5}$$

The Laplace transform of the temporal component of the solution of (218-5) gives

$$y(x,s) = A \cosh \frac{s}{a} x + B \sinh \frac{s}{a} x \tag{218-6}$$

Laplace transforming (218-2) we get $y(0,s) = 0$, which upon substituting in (218-6), gives

$$A = 0 \tag{218-6.1}$$

Thus, $\qquad y(x,s) = B \sinh \frac{s}{a} x \tag{218-6.2}$

Laplace transforming (218-3), we get

$$E y_x(1,s) = \frac{F_o}{s} \tag{218-6.3}$$

Which upon substituting in the x-derivative of (218-6.2), gives

$$E \frac{\partial y(x,s)}{\partial x} \bigg|_{x=l} = EB \frac{s}{a} \cosh \frac{s}{a} l = \frac{F_o}{s} \tag{218-6.4}$$

$$B = \frac{a F_o}{s^2.E \cosh \frac{s}{a} l} = \frac{a F_o}{E} . \frac{1}{s^2} \operatorname{sech} \frac{s}{a} l \tag{218-6.5}$$

Therefore, (218-6.2) becomes

$$y(x,s) = \frac{a F_o}{E} . \frac{1}{s^2} \operatorname{sech}\left(\frac{s}{a} l\right) \sinh\left(\frac{s}{a} x\right) \tag{218-7}$$

A $x = l$, we get

$$y(l,s) = \frac{a F_o}{E} . \frac{1}{s^2} \tanh\left(\frac{s}{a} l\right) \tag{218-7.1}$$

Now we already know that the Laplace transform of the triangular wave H(c,t) defined by

$$
\begin{aligned}
H(c,t) \quad &= t & 0 < t < c \\
&= 2c - t & c < t < 2c \\
H(c, t + 2c) \quad &= H(c,t) &
\end{aligned}
$$

is given by

$$L\left\{ H(x,t)\right\} = . \frac{1}{s^2} \tanh\left(\frac{cs}{2}\right)$$

(218-7.2)

Therefore, the inverse Laplace transform is

$$L^{-1}\left\{ \frac{aF_o}{E} \frac{1}{s^2} \tanh\frac{sl}{a}\right\} = \frac{aF_o}{E} H\left(\frac{2l}{a}, t\right)$$

(218-8)

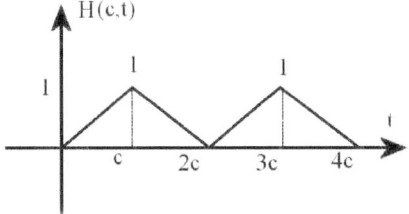

Example 148

A uniform vertical rod of mass m and length l is clamped at the end $x = 0$ and is mass-loaded by a mass M at its other end $x = l$.

Investigate the behavior of the system to an arbitrary excitation F(t) applied to the mass M.

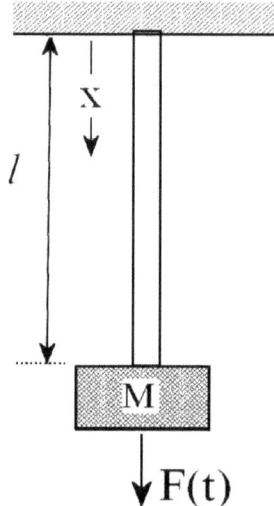

Solution

The partial differential equation satisfied by the **displacement Y** of an element at distance x from the top is

$$\frac{\partial^2 Y}{\partial t^2} = a^2 \frac{\partial^2 Y}{\partial x^2} \qquad a = \sqrt{\frac{E}{\rho}} \tag{219}$$

i.e.,
$$Y_{tt}(x,t) = a^2 Y_{xxt}(x,t) \tag{219-1}$$

and since the system is initially at rest, its transform is

$$s^2 y(x,s) = a^2 y_{xx}(x,s) \tag{219-2}$$

i.e.,
$$\frac{d^2 y(x,s)}{dx^2} - \frac{s^2}{a^2} y(x,s) = 0 \tag{219-3}$$

Therefore,

$$y(x,s) = B\cosh\frac{s}{a}x + C\sinh\frac{s}{a}x \tag{219-4}$$

From the boundary condition, Y (0,t) = 0, therefore, y(0,s) = 0, leading to B = 0.

Thus, (219-4) becomes

$$y(x,s) = C\sinh\frac{s}{a}x \qquad (219\text{-}4.1)$$

The constant C can be determined from the boundary condition at $x = l$.

$$M\left[\frac{\partial^2 Y}{\partial t^2}\right]_{x=l} = F(t) - AE\left[\frac{\partial Y}{\partial t}\right]_{x=l} \qquad (219\text{-}5)$$

Where A is the cross sectional area of the rod.

The Laplace transform of (219-5) is

$$Ms^2 y(l,s) = f(s) - AEy_x(l,s) \qquad (219\text{-}5.1)$$

Now, from (219-4), the derivative $y_x(l,s)$ is

$$y_x(x,s) = C\frac{s}{a}\cosh\frac{s}{a}x \qquad (219\text{-}5.2)$$

Thus, from (219-4.1), (219-5.1) and (219-5.2), we get

$$Ms^2 C\sinh\frac{s}{a}l = f(s) - AEC\frac{s}{a}\cosh\frac{s}{a}l \qquad (219\text{-}5.3)$$

$$C = \frac{f(s)}{Ms^2 \sinh\frac{s}{a}l + AEC\frac{s}{a}\cosh\frac{s}{a}l} \qquad (219\text{-}5.4)$$

Substituting by C, from (219-5.4), equation (219-4.1) becomes

$$y(x,s) = \frac{f(s)}{Ms^2 \sinh\frac{s}{a}l + AEC\frac{s}{a}\cosh\frac{s}{a}l}\sinh\frac{s}{a}x \qquad (219\text{-}6)$$

The inverse Laplace transform of (219-6) is

$$Y(x,t) = L^{-1}\left\{\frac{f(s)}{Ms^2 \sinh\frac{s}{a}l + AEC\frac{s}{a}\cosh\frac{s}{a}l}\sinh\frac{s}{a}x\right\}$$

$$= \frac{1}{2\pi i} \int_{\gamma-i\infty}^{\gamma+i\infty} e^{st} \frac{f(s)\sinh\dfrac{s}{a}x}{Ms^2 \sinh\dfrac{s}{a}1 + AEC\dfrac{s}{a}\cosh\dfrac{s}{a}1} .ds \qquad (219\text{-}7)$$

6.3. Partial differential equations of transmission lines

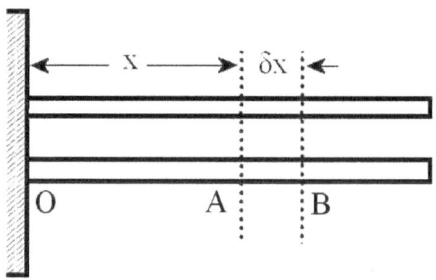

Let L,C,R,G be respectively the **inductance**, **capacitance**, **resistance** find **dielectric conductance** or leakance, all per unit length of a transmission line.

Consider an element of length δx of this line. Let the potential at A be V, then the potential at B is

$$V + \frac{\partial V}{\partial x}\delta x \qquad (220\text{-}1.1)$$

The **drop of potential** is

$$-\frac{\partial V}{\partial x}\delta x \qquad (220\text{-}1.2)$$

This is partly due to resistance R δx and partly due to inductance L δx.

Therefore,

$$IR\delta s + L\delta s\frac{\partial I}{\partial t} = -\frac{\partial V}{\partial x}\delta x$$

i.e., $\qquad \left(R + L\frac{\partial}{\partial t}\right)I = -\frac{\partial V}{\partial x} \qquad (220\text{-}1.3)$

288

Next consider an **interval of time δt.** Charge of electricity entering the element through section A in this interval is **I δt**. Charge leaving the element out of the section B in the same interval is

$$\left(I + \frac{\partial I}{\partial x}\delta x\right)\delta t \qquad\qquad (220\text{-}2.1)$$

Hence net outflow of charge from these two sections is

$$\frac{\partial I}{\partial x}\delta x \delta t \qquad\qquad (220\text{-}2.2)$$

To this we should add the charge leaking from core to sheath namely

$$G\delta x V\delta t \qquad\qquad (220\text{-}2.3)$$

Hence the total **outflow of charge** from element is

$$\frac{\partial I}{\partial x}\delta x \delta t + G\delta x V\delta t \qquad\qquad (220\text{-}2.4)$$

This will cause a **drop of potential**

$$-\frac{\partial V}{\partial t}\delta t \qquad\qquad (220\text{-}2.5)$$

Therefore,

$$\frac{\dfrac{\partial I}{\partial x}\delta x \delta t + G\delta x V\delta t}{C\delta x} = -\frac{\partial V}{\partial t}\delta t \qquad\qquad (220\text{-}2.6)$$

from which we get

$$\left(C\frac{\partial}{\partial t} + G\right)V = -\frac{\partial I}{\partial x} \qquad\qquad (220\text{-}2.7)$$

Now operating on both sides of this equation *by*

$$L\frac{\partial}{\partial t} + R \qquad\qquad (220\ 3.1)$$

we get

$$\left(L\frac{\partial}{\partial t} + R\right)\left(C\frac{\partial}{\partial t} + G\right)V = -\left(L\frac{\partial}{\partial t} + R\right)\frac{\partial I}{\partial x}$$

$$= -\frac{\partial}{\partial x}\left(L\frac{\partial}{\partial t} + R\right)I \qquad\qquad (220\text{-}3.2)$$

Substituting from (220-1.3) in (220-3.2), we get

$$\left(L\frac{\partial}{\partial t} + R \right)\left(C\frac{\partial}{\partial t} + G \right)V = \frac{\partial^2 V}{\partial x^2}$$
(220-3.3)

Therefore,

$$CL\frac{\partial^2 V}{\partial t^2} + (CR + LG)\frac{\partial V}{\partial t} + RGV = \frac{\partial^2 V}{\partial x^2}$$
(220-4)

i.e., $\quad CLV_{tt}(x,t) + (CR + LG)V_t(x,t) + RGV(x,t) = V_{xx}(x,t)$ (220-4.1)

This is the partial differential equation satisfied by **V** and we have a similar equation satisfied by **I**, namely

$$CL\frac{\partial^2 I}{\partial t^2} + (CR + LG)\frac{\partial I}{\partial t} + RGV = \frac{\partial^2 I}{\partial x^2}$$
(220-5)

i.e., $\quad CLI_{tt}(x,t) + (CR + LG)I_t(x,t) + RGI(x,t) = I_{xx}(x,t)$ (220-5.1)

Now let us apply the Laplace transformation to the above equations.

Write equations (220-13) and (220-2.7) in the form:

$$LI_t(x,t) + RI(x,t) = -V_x(x,t)$$ (220-6.1)
$$CV_t(x,t) + GV(x,t) = -I_x(x,t)$$ (220-6.2)

Laplace transforming with respect to t we get:

$$sL.i(x,s) - L.I(x,0) + R.i(x,s) = -v_x(x,s)$$ (220-7.1)
$$s.C.v(x,s) - CV(x,0) + Gv(x,s) = -i_x(x,s)$$ (220-7.2)

Therefore,

$$(sL + R)i(x,s) = L.I(x,0) - v_x(x,s)$$ (220-8.1)
$$(sC + G)v(x,s) = CV(x,0) - i_x(x,s)$$ (220-8.2)

Eliminating i(x,s) between (220-8.1) and (220-8.2) by differentiating (220-8.1) partially with respect to x and substituting for $i_x(x,s)$ from (4) we get

First, differentiating (220-8.1) with respect x

$$(sL + R)i_x(x,s) = L.I_x(x,0) - v_{xx}(x,s)$$

i.e., $\quad .i_x(x,s) = \dfrac{L.I_x(x,0) - v_{xx}(x,s)}{sL + R}$ $\hspace{3cm}$ (220-8.3)

Second, substitute by $i_x(x,s)$ in (220-8.2), we get

$$(sC + G)v(x,s) = CV(x,0) - \dfrac{L.I_x(x,0) - v_{xx}(x,s)}{sL + R} \hspace{2cm} (220\text{-}8.4)$$

$$v_{xx}(x,s) - (sL + R)(sC + G)v(x,s) = L.I_x(x,0) - (sL + R)CV(x,0) \hspace{1cm} (220\text{-}9)$$

$$\dfrac{d^2 v(x,s)}{dx^2} - q^2 v(x,s) = L\dfrac{dI(x,0)}{dx} - C(sL + R)V(x,0) \hspace{2cm} (220\text{-}9.1)$$

Where,

$$q^2 = (Ls + R)(Cs + G) \hspace{5cm} (220\text{-}9.2)$$

Now, suppose we have a semi— infinite line $x \geq o$ with the following conditions:

(i) Initial currents and charges are zeros. $\hspace{3cm}$ (220-9.2.1)
(ii) $\hspace{2cm} \lim_{x \to \infty} V(x,t) = 0$ $\hspace{3cm}$ (220-9.2.2)
(iii) $\hspace{2cm} V(0,t) = F(t)$ $\hspace{3.5cm}$ (220-9.2.3)

Those initial conditions lead to

(ii.a) $\hspace{2cm} \lim_{x \to \infty} v(x,s) = 0$ $\hspace{3cm}$ (220-9.2.4)
(iii.a) $\hspace{2cm} v(0,s) = f(s)$ $\hspace{3.5cm}$ (220-9.2.5)

Hence the differential equation (220-9.1), in v(x,s), reduces to

$$\dfrac{d^2 v(x,s)}{dx^2} - q^2 v(x,s) = 0 \hspace{4cm} (220\text{-}9.3)$$

Subject to the initial conditions in (220-9.2.1) through (220-9.2.5), and equation (220-9.3) has the solution

$$v(x,s) = A\,e^{qx} + B\,e^{-qx} \hspace{4cm} (220\text{-}10)$$

From (220-9.2.4), $v(\infty,s) = 0$, then $A = 0$
And, from (220-9.2.5), since $v(0,s) = f(s)$ then $f(s) - B$

Hence,

$$v(x,s) = f(s)\,e^{-qx} \hspace{4.5cm} (220\text{-}10.1)$$

From (220-9.2), we get

i.e., $\qquad v(x,s) = f(s)e^{-\sqrt{(Ls+R)(Cs+G)}.x}$ (220-10.2)

The inverse transform in the general case is a complicated one. The following special but important cases will now be investigated.

Case I: **The lossless line R = 0 , G =0**

Here, equation (220-10.2) reduces to

$$v(x,s) = f(s)e^{-\sqrt{LC}.sx}$$ (221)

Put $\qquad u = \dfrac{1}{\sqrt{LC}}$ (221-1)

Therefore,

$$v(x,s) = f(s)e^{-\frac{x}{u}s}$$ (221-1.1)

The inverse Laplace transform of (220-11.2) is a step function with period equal to x / u.

Therefore,

$$V(x,t) = L^{-1}\left\{f(s)e^{-\frac{x}{u}s}\right\} = U\left(t - \frac{x}{u}\right).F\left(t - \frac{x}{u}\right)$$ (221-2)

This means that a point at distance x from the origin remains at zero potential until a time $t = \dfrac{x}{u}$ elapses when the voltage F(t), at x = 0, is transferred to the point. The time x/u is the time necessary for a **wave traveling** with **velocity u** to describe the distance x, where
$$u = \frac{1}{\sqrt{LC}}$$

Case II: **The Heaviside's distortionless line CR = LG**

Starting from equation (220-9.2), put

$$\frac{R}{L} = \frac{G}{C} = \lambda$$ (222)

Therefore, equation (220-9.2) becomes

$$q^2 = (Ls + R)(Cs + G) = (Ls + \lambda L)(Cs + \lambda C) = CL(s + \lambda)^2$$ (222-1)

Therefore,

292

$$q = \sqrt{CL}(s+\lambda) = \frac{s+\lambda}{u} \qquad (222\text{-}2)$$

Where, again $\quad u = \dfrac{1}{\sqrt{LC}} \qquad (222\text{-}2.1)$

The inverse Laplace transform of (220-10.1) is, therefore,

$$V(x,t) = L^{-1}\left\{ f(s)e^{-\frac{s+\lambda}{u}x} \right\} = e^{-\frac{\lambda x}{u}} L^{-1}\left\{ f(s)e^{-\frac{s}{u}x} \right\}$$

$$= e^{-\frac{\lambda x}{u}} U\!\left(t - \frac{x}{u} \right)\!.F\!\left(t - \frac{x}{u} \right) \qquad (222\text{-}3)$$

Hence, we have a solution similar to the previous case but with **attenuation** as we move along the line due to the existence of the factor

$$e^{-\frac{\lambda x}{u}}$$

Case III: **The ideal submarine cable L = 0, G = 0**

Starting from equation (220-9.2), put **L = 0, G = 0**

$$q^2 = RCs \qquad (223)$$

Put $\qquad k = \dfrac{1}{RC} \qquad (223\text{-}1)$

Therefore, $\qquad q = \sqrt{\dfrac{s}{k}} \qquad (223\text{-}1.1)$

Hence,

$$v(x,s) = f(s)e^{-\sqrt{\frac{s}{k}}x} \qquad (223\text{-}2)$$

Suppose that $F(t) = E$, where E is constant, then

$$f(s) = E / s \qquad (223\text{-}2.1)$$

$$v(x,s) = \frac{E}{s}e^{-\sqrt{\frac{s}{k}}x} \qquad (223\text{-}2.2)$$

Now, we have shown earlier that

$$L^{-1}\left\{ \frac{1}{s}e^{-\sqrt{\frac{s}{k}}x} \right\} = \text{erfc}\frac{x}{2\sqrt{kt}} \qquad (223\text{-}3)$$

Therefore, the inverse Laplace transform of (223-2.2) is

$$V(x,t) = L^{-1}\left\{\frac{E}{s}e^{-\sqrt{\frac{s}{k}}x}\right\} = E \, erfc\frac{x}{2\sqrt{kt}} \qquad (223\text{-}4)$$

This problem, as we shall see, has an analogous problem in **linear heat flow**.

6.4. Conduction of heat

Let ρ = **density** of conducting medium, **c** its **specific heat** and K its **thermal conductivity**. Consider an infinitesimal rectangular parallelepiped, whose centre P is the point **P(x,y,z)** and whose faces are parallel to the coordinate planes and let the lengths of its sides be **δx, δy,** and **δz.**

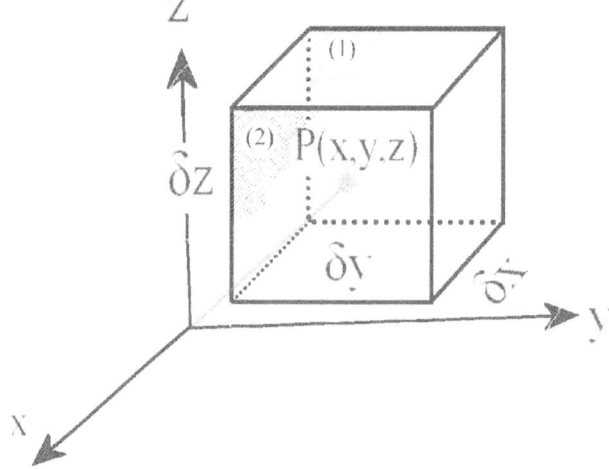

The quantity of heat entering the element in time **δt** through the face (1) is

$$-K\delta x\delta y\left(\frac{\partial U(x,y,z,t)}{\partial x} - \frac{\partial^2 U(x,y,z,t)}{\partial x^2}\cdot\frac{\delta x}{2}\right)\delta t \qquad (224\text{-}1.1)$$

The quantity of heat leaving the element in time **δt** from face (2) is

$$-K\delta x\delta y\left(\frac{\partial U(x,y,z,t)}{\partial x} + \frac{\partial^2 U(x,y,z,t)}{\partial x^2}\cdot\frac{\delta x}{2}\right)\delta t \qquad (224\text{-}1.2)$$

Hence net gain of heat from the two faces is

$$K\delta x \delta y . \delta t \frac{\partial^2 U(x, y, z, t)}{\partial x^2} \qquad (224\text{-}1.3)$$

We have two similar contributions from the remaining two pairs of parallel faces. Hence total gain of heat in the element is

$$K\delta x \delta y \delta z . \delta t \left(\frac{\partial^2 U(x, y, z, t)}{\partial x^2} + \frac{\partial^2 U(x, y, z, t)}{\partial y^2} + \frac{\partial^2 U(x, y, z, t)}{\partial z^2} \right) \qquad (224\text{-}1.4)$$

This will increase the temperature U(x,y,z,t) by

$$\rho . c . \frac{\partial U(x, y, z, t)}{\partial t} \delta t \qquad (224\text{-}1.5)$$

Therefore, the balance heat flow versus temperature rise of internal matter in the volume element is

$$K\delta x \delta y \delta z . \delta t \left(\frac{\partial^2 U(x, y, z, t)}{\partial x^2} + \frac{\partial^2 U(x, y, z, t)}{\partial y^2} + \frac{\partial^2 U(x, y, z, t)}{\partial z^2} \right) = \rho . c . \delta x \delta y \delta z . \frac{\partial U(x, y, z, t)}{\partial t} \delta t$$

i.e.,

$$\frac{\partial^2 U(x, y, z, t)}{\partial x^2} + \frac{\partial^2 U(x, y, z, t)}{\partial y^2} + \frac{\partial^2 U(x, y, z, t)}{\partial z^2} = \frac{\rho . c}{K} . \frac{\partial U(x, y, z, t)}{\partial t} \qquad (224\text{-}2)$$

$$\nabla^2 U(x, y, z, t) = \frac{\rho . c}{K} . \frac{\partial U(x, y, z, t)}{\partial t} \qquad (224\text{-}3)$$

Put
$$k = \frac{K}{\rho . c} \qquad (224\text{-}3.1)$$

Where k is known as the **thermal diffusivity** of the material.
∇^2 is **Laplacian operator** defined in the Cartesian coordinates by

$$\nabla^2 \cong \frac{\partial^2}{\partial x^2} + \frac{\partial^2}{\partial y^2} + \frac{\partial^2}{\partial z^2} \qquad (224\text{-}3.1.1)$$

Therefore,

$$\frac{\partial U}{\partial t} = k \nabla^2 U \qquad (224\text{-}3.2)$$

This is known as the **heat equation or equation of diffusion**.

If the flow of heat is **uniflow** i.e., in one direction, say the x - axis then the last equation becomes

$$\frac{\partial U}{\partial t} = k\frac{\partial^2 U}{\partial x^2} \qquad\qquad (224\text{-}3.3)$$

When the flow of heat is **steady** i.e., independent of the time, the heat equation reduces to

$$\nabla^2 U = 0 \qquad\qquad (224\text{-}3.4)$$

i.e., the temperature U satisfies **Laplace's equation**.

Example 149

A semi-infinite solid $x \geq 0$ is initially at zero temperature. A constant temperature $U_o > 0$ is applied at time $t = 0$ to the face $x = 0$ and is maintained.

(i) Find the **temperature distribution** at any point x of the solid and at any instant, t.
(ii) Then assume that $U(0,t) = V(t)$ and find the **temperature distribution** at any point x of the solid and at any instant, t.

Solution

(i) $U(0,t) = U_o$

Here we have to solve the heat equation

$$\frac{\partial U}{\partial t} = k\frac{\partial^2 U}{\partial x^2} \qquad\qquad x > 0, \qquad t > 0 \qquad\qquad (225\text{-}1)$$

subject to the boundary conditions

$$U(x,0) = 0 \qquad\qquad (225\text{-}1.1)$$
$$U(0,t) = U_o \qquad\qquad (225\text{-}1.2)$$

Laplace transforming the differential equation (225-1) with respect to t, we get

$$su(x,s) - U(x,0) = ku_{xx}(x,s) \qquad\qquad (225\text{-}2)$$

i.e.,
$$\frac{d^2 u(x,s)}{dx^2} - \frac{s}{k}u(x,s) = 0 \qquad\qquad (225\text{-}2.1)$$

With the solution

$$u(x,s) = Ae^{\sqrt{\frac{s}{k}}x} + Be^{-\sqrt{\frac{s}{k}}x} \qquad\qquad (225\text{-}3)$$

Since U(x,t) is bounded and consequently u(x,s) is also bounded, then A =0 and

$$u(x,s) = Be^{-\sqrt{\frac{s}{k}}x} \tag{225-3.1}$$

Since U(0,t) = U$_o$, then Laplace transforming U(0,t), we get u(0,s) = U$_o$ /s and

$$B = U_o /s$$

Therefore,

$$u(x,s) = \frac{U_o}{s} e^{-\sqrt{\frac{s}{k}}x} \tag{225-3.2}$$

The inverse Laplace transform of (225-3.2)

$$U(x,t) = L^{-1}\left\{\frac{U_o}{s} e^{-\sqrt{\frac{s}{k}}x}\right\} = U_o erfc\left(\frac{x}{2\sqrt{kt}}\right) \tag{225-4}$$

(ii) U(0,t) = V(t)

Laplace transforming U(0,t), we get

$$u(0,s) = v(s)$$
$$B = v(s) \tag{226-1}$$

Therefore, (225-3.1) becomes

$$u(x,s) = v(s)e^{-\sqrt{\frac{s}{k}}x} \tag{226-2}$$

Now, we could write

$$L^{-1}\left\{e^{-\sqrt{\frac{s}{k}}x}\right\} = \frac{x}{2\sqrt{\pi k}} t^{-\frac{3}{2}} e^{-\frac{x^2}{4kt}} \tag{226-3}$$

Hence according to a theorem on convolution we have

$$U(x,t) = \frac{x}{2\sqrt{\pi k}} \int_0^t V(t-\lambda)\lambda^{-\frac{3}{2}} e^{-\frac{x^2}{4k\lambda}} d\lambda \tag{226-4}$$

Example 150

A solid is bounded by the infinite planes x = 0 and x = 1.

These planes are kept at zero temperature. If at time t = 0, the distribution of temperature is

$$U(x,0) = 4 \sin 2 \pi x \qquad (227)$$

Find the temperature at any point of the solid and at any instant, t.

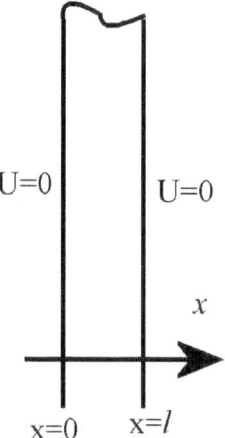

U=0   U=0

x

x=0   x=l

Solution

The governing heat equation is

$$\frac{\partial U}{\partial t} = k \frac{\partial^2 U}{\partial x^2} \qquad x > 0 , \qquad t > 0 \qquad (227\text{-}1)$$

subject to the boundary and initial conditions

$$U(0,t) = 0 \qquad\qquad\qquad (227\text{-}1.1)$$
$$U(l,t) = 0 \qquad\qquad\qquad (227\text{-}1.2)$$
$$U(x,0) = 4 \sin 2 \pi x \qquad\qquad (227\text{-}1.3)$$

$$su(x,s) - U(x,0) = ku_{xx}(x,s) \qquad (227\text{-}2)$$

i.e.,
$$\frac{d^2 u(x,s)}{dx^2} - \frac{s}{k} u(x,s) = -\frac{4}{k} \sin 2\pi x \qquad (227\text{-}2.1)$$

With the solution

$$u(x,s) = A e^{\sqrt{\frac{s}{k}}x} + B e^{-\sqrt{\frac{s}{k}}x} + \frac{4}{s + 4\pi^2 k} \sin 2\pi x \qquad (227\text{-}3)$$

The boundary condition $U(0,t) = 0$, gives $u(0,s) = 0$, therefore

$$0 = A + B, \text{ or } A = -B \tag{227-3.1}$$

The boundary condition $U(l,t) = 0$, gives $u(l,s) = 0$, and since $A = -B$, therefore

$$0 = -Be^{\sqrt{\frac{s}{k}}} + Be^{-\sqrt{\frac{s}{k}}} + \frac{4}{s + 4\pi^2 k} \sin 2\pi$$

$$0 = -Be^{\sqrt{\frac{s}{k}}} + Be^{-\sqrt{\frac{s}{k}}} + \frac{4}{s + 4\pi^2 k} 0 \tag{227-3.2}$$

Therefore, $A = B = 0$

Thus, (227-3) becomes

$$u(x,s) = \frac{4}{s + 4\pi^2 k} \sin 2\pi x \tag{227-3.3}$$

i.e.,

$$U(x,t) = 4\sin(2\pi x) L^{-1}\left\{ \frac{1}{s + 4\pi^2 k} \right\} = 4e^{-4\pi^2 kt} \sin(2\pi x) \tag{227-4}$$

Example 151

A semi-infinite insulated bar is initially at zero temperature. The bar is placed along the positive x -.axis. At $t = 0$, a quantity of heat is instantaneously generated at the point $x = a$ where a is positive. Find the **temperature distribution** along the bar at any instant.

Solution

The partial differential equation satisfied by the temperature U *is*

$$\frac{\partial U}{\partial t} = k\frac{\partial^2 U}{\partial x^2} \qquad\qquad x > 0, \qquad\qquad t > 0 \tag{228}$$

The boundary conditions are

$$U(x,0) = 0 \tag{228-1.1}$$
$$U(x,t) \text{ is bounded} \tag{228-1.2}$$
$$U(a,t) = Q\delta(t) \tag{228-1.3}$$

Where Q is a constant and $\delta(t$ is the Dirac Delta function.

Therefore, $\quad su(x,s) - U(x,0) = ku_{xx}(x,s) \tag{228-2}$

i.e., $\qquad \dfrac{d^2u(x,s)}{dx^2} - \dfrac{s}{k}u(x,s) = 0$ $\hfill$ (228-2.1)

With the solution

$$u(x,s) = Ae^{\sqrt{\frac{s}{k}}x} + Be^{-\sqrt{\frac{s}{k}}x} \hfill (228\text{-}3)$$

From the boundary condition A = 0.

Therefore, $\qquad u(x,s) = Be^{-\sqrt{\frac{s}{k}}x}$ $\hfill$ (228-3.1)

Transforming U(a,t) = Qδ(t), we get

$$u(a,s) = Q = Be^{-\sqrt{\frac{s}{k}}a}$$

$$B = Qe^{\sqrt{\frac{s}{k}}a} \hfill (228\text{-}3.2)$$

Therefore,

$$u(x,s) = Qe^{-\sqrt{\frac{s}{k}}(x-a)} \hfill (228\text{-}3.3)$$

Now,

$$L^{-1}\left\{ e^{-a\sqrt{s}} \right\} = \dfrac{a}{2\sqrt{\pi t^3}} e^{-\frac{a^2}{4t}}, \qquad a > 0 \hfill (228\text{-}3.4)$$

Therefore, (228-3.3) becomes

$$U(x,t) = Q\dfrac{x-a}{2\sqrt{\pi t^3}} e^{-\frac{(a-x)^2}{4kt}} \hfill (228\text{-}4)$$

Example 152

A semi-infinite solid x ≥ 0 with zero initial temperature has a constant heat flux C applied to the face x = 0 so that

$$-KU_x(0,t) = C$$

Find the temperature at any instant t and at any point x of the solid.

Solution

$$\dfrac{\partial U}{\partial t} = k\dfrac{\partial^2 U}{\partial x^2} \qquad\qquad x > 0, \qquad t > 0 \hfill (229)$$

The boundary conditions are

$$U(x,0) = 0 \qquad\qquad\qquad (229\text{-}1.1)$$
$$-KU_x(0,t) = C \qquad\qquad\qquad (229\text{-}1.2)$$

Therefore, $\quad su(x,s) - U(x,0) = ku_{xx}(x,s) \qquad\qquad\qquad (229\text{-}2)$

i.e., $\quad \dfrac{d^2 u(x,s)}{dx^2} - \dfrac{s}{k} u(x,s) = 0 \qquad\qquad\qquad (229\text{-}2.1)$

With the solution

$$u(x,s) = Ae^{\sqrt{\frac{s}{k}}x} + Be^{-\sqrt{\frac{s}{k}}x} \qquad\qquad\qquad (229\text{-}3)$$

From the boundary condition $A = 0$.

Therefore, $\quad u(x,s) = Be^{-\sqrt{\frac{s}{k}}x} \qquad\qquad\qquad (229\text{-}3.1)$

Transforming the boundary condition $-K\,U_x(0,t) = C$, we get

$$-K\,u_x(0,s) = C/s$$
$$u_x(0,s) = -C/Ks \qquad\qquad\qquad (229\text{-}3.2)$$

From the derivative of (229-3.1), and substituting from (229-3.2), we get

$$u_x(x,s) = \frac{d}{dx}u(x,s) = -B\sqrt{\frac{s}{k}}e^{-\sqrt{\frac{s}{k}}a}$$

$$u_x(0,s) = -B\sqrt{\frac{s}{k}} = -\frac{C}{Ks}$$

$$B = \frac{C\sqrt{k}}{Ks^{\frac{3}{2}}} \qquad\qquad\qquad (229\text{-}3.3)$$

Therefore,

$$u(x,s) = \frac{C\sqrt{k}}{Ks^{\frac{3}{2}}}e^{-\sqrt{\frac{s}{k}}x} \qquad\qquad\qquad (229\text{-}4)$$

Now,

$$L^{-1}\left\{ s^{-\frac{3}{2}}e^{-a\sqrt{s}} \right\} = 2\sqrt{\frac{t}{\pi}}e^{-\frac{a^2}{4t}} - a.\mathrm{erfc}\left(\frac{a}{2\sqrt{t}}\right), \qquad a > 0 \qquad\qquad\qquad (229\text{-}4.1)$$

Therefore, (229-4) becomes

$$U(x,t) = \frac{C\sqrt{k}}{K}\left[2\sqrt{\frac{t}{\pi}}e^{-\frac{x^2}{4kt}} - \frac{x}{\sqrt{k}}.\text{erfc}\left(\frac{x}{2\sqrt{kt}}\right)\right] \qquad (229\text{-}5)$$

$$U(x,t) = \frac{C}{K}\left[2\sqrt{\frac{kt}{\pi}}e^{-\frac{x^2}{4kt}} - x.\text{erfc}\left(\frac{x}{2\sqrt{kt}}\right)\right] \qquad (229\text{-}5.1)$$

Example 153

A bar of length l is initially at temperature U_o. The end $x = o$ is insulated while the end $x = l$ is suddenly given the **constant temperature** U_1. Find the temperature at any point x of the bar and at any instant assuming the surface of the bar to be insulated.

Solution

$$\frac{\partial U}{\partial t} = k\frac{\partial^2 U}{\partial x^2} \qquad\qquad 0 < x < 10, \qquad t > 0 \qquad (230)$$

Boundary conditions are

$$U(x,0) = U_o \qquad\qquad (230\text{-}1.1)$$
$$U(l,t) = U_1 \qquad\qquad (230\text{-}1.2)$$
$$U_x(0,t) = 0 \qquad\qquad (230\text{-}1.3)$$

Therefore, $\quad su(x,s) - U(x,0) = ku_{xx}(x,s) \qquad (230\text{-}2)$

i.e., $\qquad \dfrac{d^2u(x,s)}{dx^2} - \dfrac{s}{k}u(x,s) = -\dfrac{U_o}{k} \qquad (230\text{-}2.1)$

With the solution

$$u(x,s) = A\cosh\sqrt{\frac{s}{k}}x + B\sinh\sqrt{\frac{s}{k}}x + \frac{U_o}{s} \qquad (230\text{-}3)$$

Transforming the boundary condition $U_x(0,t) = 0$, we get $u_x(0,s) = 0$, therefore

$$B = 0 \qquad (230\text{-}3.1)$$

i.e., $\qquad u(x,s) = A\cosh\sqrt{\dfrac{s}{k}}x + \dfrac{U_o}{s} \qquad (230\text{-}3.2)$

Transforming the boundary condition $U(l,t) = U_1$, we get $u(l,s) = U_1/s$, therefore

$$u(l,s) = \frac{U_1}{s} = A\cosh\sqrt{\frac{s}{k}}l + \frac{U_o}{s} \qquad (230\text{-}3.2)$$

$$A = \frac{U_1 - U_o}{s.\cosh\sqrt{\frac{s}{k}}l} \tag{230-3.3}$$

Thus, equation (230-3.2) becomes

$$u(x,s) = \frac{U_1 - U_o}{s.\cosh\sqrt{\frac{s}{k}}l}\cosh\sqrt{\frac{s}{k}}x + \frac{U_o}{s} \tag{230-3.4}$$

The Laplace inversion of (230-3.4) is

$$U(x,t) = (U_1 - U_o)L^{-1}\left\{\frac{\cosh\sqrt{\frac{s}{k}}x}{s.\cosh\sqrt{\frac{s}{k}}l}\right\} + L^{-1}\left\{\frac{U_o}{s}\right\} \tag{230-4}$$

$$U(x,t) = U_o + (U_1 - U_o)L^{-1}\left\{\frac{\cosh\sqrt{s/k}x}{s.\cosh\sqrt{s/k}l}\right\} \tag{230-4.1}$$

According to the theory of the inversion integral we have

$$L^{-1}\left\{\frac{\cosh\sqrt{s/k}x}{s.\cosh\sqrt{s/k}l}\right\} = \text{sum of residue of } \frac{e^{st}\cosh\sqrt{s/k}x}{s.\cosh\sqrt{s/k}l} \tag{230-4.2}$$

At its poles.

The poles are the zeros of $s.\cosh\sqrt{s/k}l$, i.e., $s = 0$ and the zeros of $\cosh\sqrt{s/k}l$

Now, consider the zeros of the $\cosh u = 0$

Therefore,

$$\cosh u = \frac{e^u + e^{-u}}{2} = 0 \tag{230-4.3}$$

Thus, $e^u = -e^{-u}$

$$e^{2u} = -1 = e^{\pi i + 2n\pi i}, \qquad \text{where } n = 0,1,2,3, .. \tag{230-4.4}$$

Therefore,

$$2u = (2n+1)\pi i$$

i.e., $u = (n+1/2)\pi i$ (230-4.5)

Therefore, $\cosh\sqrt{s/k}l = 0$ if

303

$$\sqrt{s/kl} = (n+1/2)\,\pi i \qquad\qquad (230\text{-}4.6)$$

which can also be written

$$s = -k\left(\frac{(2n-1)\pi}{2l}\right)^2, \qquad \text{where } n = 1,2,3,\,.. \qquad\qquad (230\text{-}4.7)$$

Where we changed the initial value of n from zero to 1. Hence the zeros of $s.\cosh\sqrt{s/kl}$ are determined by $s = 0$ and by equations (230-4.7).

Determining the residues of (230-4.2):

(i) Residue at s = 0

$$\lim_{s \to 0} s\,\frac{e^{st}\cosh\sqrt{s/kx}}{s.\cosh\sqrt{s/kl}} = 1 \qquad\qquad (230\text{-}4.8)$$

(ii) Residue at $s = -k\left(\dfrac{(2n-1)\pi}{2l}\right)^2$ **,where n =1,2,3, ..**

$$\lim_{s \to s_n}\left(s-s_n\right)\frac{e^{st}\cosh\sqrt{s/kx}}{s.\cosh\sqrt{s/kl}} = \left[\lim_{s \to s_n}\frac{s-s_n}{\cosh\sqrt{s/kl}}\right]\left[\lim_{s \to s_n}\frac{e^{st}\cosh\sqrt{s/kx}}{s}\right] \qquad (230\text{-}4.9)$$

$$= \left[\lim_{s \to s_n}\frac{1}{\sinh\left(\sqrt{s/kl}\right)\left(\dfrac{1}{2\sqrt{ks}}\right)}\right]\left[\lim_{s \to s_n}\frac{e^{st}\cosh\sqrt{s/kx}}{s}\right]$$

$$= \frac{4(-1)^n}{(2n-1)\pi}e^{-kt\left(\frac{(2n-1)\pi}{2l}\right)^2}\cos\frac{(2n-1)\pi}{2l}x \qquad\qquad (230\text{-}4.10)$$

According to **L'Hiospital's rule** in evaluating indeterminate forms, and the fact that

$$\cosh iu\; \cos u \qquad\qquad (230\text{-}4.11)$$
$$\sinh iu = i\sin u \qquad\qquad (230\text{-}4.12)$$

Therefore, equation (230-4.1) becomes

$$U(x,t) = U_1 + \frac{4\left(U_1 - U_o\right)}{\pi} \sum_{n=1}^{\infty} \frac{(-1)^n}{(2n-1)} e^{-kt\left(\frac{(2n-1)\pi}{2l}\right)^2} \cos\frac{(2n-1)\pi}{2l}x \qquad (230\text{-}5)$$

Example 154

An infinitely long circular cylinder of radius unity is initially at temperature U_o. At time $t = 0$, a temperature $0°C$ is applied to the surface and is maintained. Determine the temperature distribution over the cylinder at any instant t.

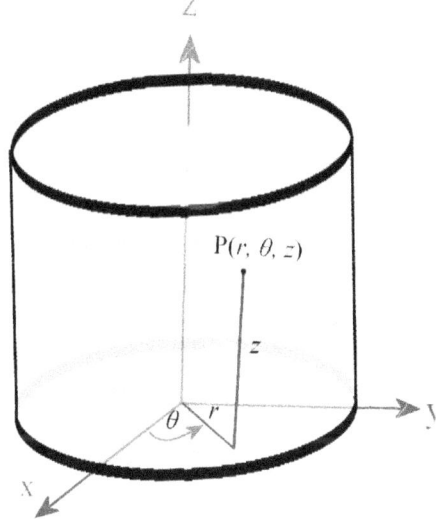

Solution

The equation satisfied by the temperature U is

$$\frac{\partial U}{\partial t} = k\nabla^2 U \qquad (231)$$

Now in cylindrical coordinates the **Laplacian operator** takes the form

$$\nabla^2 \cong \frac{\partial^2}{\partial r^2} + \frac{1}{r}\cdot\frac{\partial}{\partial r} + \frac{1}{r^2}\cdot\frac{\partial^2}{\partial \theta^2} + \frac{\partial^2}{\partial z^2} \qquad (231\text{-}1)$$

and hence the heat equation (231) becomes

$$\frac{\partial U}{\partial t} = k\left[\frac{\partial^2 U}{\partial r^2} + \frac{1}{r}\cdot\frac{\partial U}{\partial r} + \frac{1}{r^2}\cdot\frac{\partial^2 U}{\partial \theta^2} + \frac{\partial^2 U}{\partial z^2}\right] \qquad (231\text{-}2)$$

Where,

$$U = U\,(r,\,\theta,\,z) \qquad (231\text{-}2.1)$$

Since in this problem, due to symmetry, it is independent of θ and z, then

$$\frac{\partial U}{\partial t} = k\left(\frac{\partial^2 U}{\partial r^2} + \frac{1}{r}\cdot\frac{\partial U}{\partial r}\right), \qquad 0 < r < 1 \qquad (231\text{-}3)$$

Now, it is more convenient to take this equation in the form

$$\frac{\partial U}{\partial t} = \frac{\partial^2 U}{\partial r^2} + \frac{1}{r}\cdot\frac{\partial U}{\partial r} \qquad (231\text{-}3.1)$$

and then, in the final result, t is changed to kt.

The boundary conditions are
$U(r,0) = U_o$ (231-4.1)
$U(l,t) = 0$ (231-4.2)
$U(r,t)$ is bounded   (231-4.3)

Laplace transform the heat equation (231-3.1) with respect to t, we get

$$su(r,s) - U(r,0) = \frac{d^2 u(r,s)}{dr^2} + \frac{1}{r}\cdot\frac{du(r,s)}{dr} \qquad (231\text{-}5)$$

i.e., $\dfrac{d^2 u(x,s)}{dr^2} + \dfrac{1}{r}\cdot\dfrac{du(x,s)}{dr} - su(x,s) = -U_o \qquad (231\text{-}5.1)$

With the solution

$$u(r,s) = AJ_o\left(i\sqrt{s}\,r\right) + BY_o\left(i\sqrt{s}\,r\right) + \frac{U_o}{s} \qquad (231\text{-}6)$$

Since u(r,s) is finite then B = 0

$$u(r,s) = AJ_o\left(i\sqrt{s}\,r\right) + \frac{U_o}{s} \qquad (231\text{-}6.1)$$

Laplace transforming $U(l,t) = 0$, from (231-4.2), we get u(l,s) = 0.

Therefore,

$$u(l,s) = 0 = AJ_o\left(i\sqrt{s}\,l\right) + \frac{U_o}{s}$$

i.e., $\qquad A = -\dfrac{U_{_o}}{sJ_{_o}\left(i\sqrt{s}\right)}$ (231-6.2)

Therefore, the Laplace transform of the heat equation (231-6) is determined as

$$u(r,s) = \dfrac{U_{_o}}{s} - \dfrac{U_{_o}}{sJ_{_o}\left(i\sqrt{s}\right)}J_{_o}\left(i\sqrt{s}\,r\right)$$ (231-6.3)

The inverse Laplace transform of (231-6.3) is

$$U(r,t) = L^{-1}\left\{\dfrac{U_{_o}}{s} - \dfrac{U_{_o}}{sJ_{_o}\left(i\sqrt{s}\right)}J_{_o}\left(i\sqrt{s}\,r\right)\right\}$$

$$= U_{_o} - U_{_o}L^{-1}\left\{\dfrac{J_{_o}\left(i\sqrt{s}\,r\right)}{sJ_{_o}\left(i\sqrt{s}\right)}\right\}$$ (231-7)

Now according to the **inversion integral formula**

$$L^{-1}\left\{\dfrac{J_{_o}\left(i\sqrt{s}\,r\right)}{sJ_{_o}\left(i\sqrt{s}\right)}\right\} = \text{sum of residues of } \dfrac{e^{st}J_{_o}\left(i\sqrt{s}\,r\right)}{sJ_{_o}\left(i\sqrt{s}\right)} \text{ at its poles}$$

The poles are the zeros of $s.J_{_o}\left(i\sqrt{s}\right)$, i.e., $s = 0$ and the zeros of $J_{_o}\left(i\sqrt{s}\right)$.
Now, $J_{_o}\left(i\sqrt{s}\right)$ has simple zeros at

$$i\sqrt{s} = \lambda_{_1}, \lambda_{_2}, ..., \lambda_{_n},$$ (231-7.1)

i.e., $\qquad s = -\lambda_{_1}^{2}, -\lambda_{_2}^{2}, ..., -\lambda_{_n}^{2},$ (231-7.2)

Determination of residues:

(i) Residue at $s = 0$ is $\qquad \lim\limits_{s \to 0}\dfrac{e^{st}J_{_o}\left(i\sqrt{s}\,r\right)}{sJ_{_o}\left(i\sqrt{s}\right)} = 1$ (231-7.3)

(ii) Residues at $s = -\lambda_{_1}^{2}, -\lambda_{_2}^{2}, ..., -\lambda_{_n}^{2},$ are

$$\lim\limits_{s \to -\lambda_{_n}^{2}}\left(s + \lambda_{_n}^{2}\right)\dfrac{e^{st}J_{_o}\left(i\sqrt{s}\,r\right)}{sJ_{_o}\left(i\sqrt{s}\right)} = \left[\lim\limits_{s \to -\lambda_{_n}^{2}}\dfrac{s + \lambda_{_n}^{2}}{J_{_o}\left(i\sqrt{s}\right)}\right]\left[\lim\limits_{s \to -\lambda_{_n}^{2}}\dfrac{e^{st}J_{_o}\left(i\sqrt{s}\,r\right)}{s}\right]$$ (231-7.4)

$$= \left[\lim_{s \to -\lambda_n^2} \frac{1}{J'_o(i\sqrt{s})\dfrac{i}{2\sqrt{s}}} \right] \left[\frac{e^{-\lambda_n^2 t} J_o(\lambda_n r)}{-\lambda_n^2} \right] \qquad (231\text{-}7.5)$$

$$= -\frac{2e^{-\lambda_n^2 t} J_o(\lambda_n r)}{\lambda_n J_1(\lambda_n)} \qquad (231\text{-}7.6)$$

According to **L' Hospital's rule** in evaluating indeterminate forms and the fact that $J'_o(x) = -J_1(x)$, the inverse Laplace transform of (231-7) is

$$U(r,t) = U_o - U_o \left[1 - 2\sum_{n=1}^{\infty} \frac{e^{-\lambda_n^2 t} J_o(\lambda_n r)}{\lambda_n J_1(\lambda_n)} \right] \qquad (231\text{-}8)$$

Changing t into kt we get

$$U(r,t) = 2U_o \sum_{n=1}^{\infty} \frac{e^{-k\lambda_n^2 t} J_o(\lambda_n r)}{\lambda_n J_1(\lambda_n)} \qquad (231\text{-}8.1)$$

Example 155

A semi-infinite solid $x \geq 0$ has zero initial temperature.

The temperature at $x = 0$ is given by

$$U(0,t) \qquad = U_o \qquad 0 < t < t_o \qquad (232\text{-}1.1)$$
$$= 0 \qquad t < t_o \qquad (232\text{-}1.2)$$

Determine the temperature at any point x of the solid and at any instant t.

Solution

$$\frac{\partial U}{\partial t} = k\frac{\partial^2 U}{\partial x^2} \qquad (232\text{-}2)$$

Laplace transforming the differential equation (232-2) with respect to t, we get

$$su(x,s) - U(x,0) = ku_{xx}(x,s) \qquad (232\text{-}3)$$

i.e., $$\frac{d^2 u(x,s)}{dx^2} - \frac{s}{k}u(x,s) = 0 \qquad (232\text{-}3.1)$$

With the solution

$$u(x,s) = Ae^{\sqrt{\frac{s}{k}}x} + Be^{-\sqrt{\frac{s}{k}}x} \qquad\qquad (232\text{-}4)$$

Since U(x,t) is bounded and consequently u(x,s) is also bounded, then A =0 and

$$u(x,s) = Be^{-\sqrt{\frac{s}{k}}x} \qquad\qquad (232\text{-}4.1)$$

Since U(0,t) is step function defined by (232-1.1) and (232-1.2) then Laplace transforming U(0,t), we get

$$u(0,s) = \frac{U_o}{s}\left(1 - e^{-t_o s}\right) \qquad\qquad (232\text{-}4.2)$$

Therefore, $\qquad B = \dfrac{U_o}{s}\left(1 - e^{-t_o s}\right) \qquad\qquad (232\text{-}4.3)$

Thus, equation (232-4.1) becomes

$$u(x,s) = \frac{U_o}{s}\left(1 - e^{-t_o s}\right)e^{-\sqrt{\frac{s}{k}}x} \qquad\qquad (232\text{-}4.4)$$

Now, the inverse Laplace transform of the first part of (232-4.4), gives

$$L^{-1}\left\{\frac{1}{s}e^{-a\sqrt{s}}\right\} = \text{erfc}\left(\frac{a}{2\sqrt{t}}\right), \qquad\qquad a > 0 \qquad\qquad (232\text{-}5.1)$$

Therefore, the inverse Laplace transform of (232-4.4) consists of the that of (232-5.1) in addition to the step unit function term.

$$U(x,t) = U_o\left[\text{erfc}\left(\frac{x}{2\sqrt{kt}}\right) - U(t - t_o).\text{erfc}\left(\frac{x}{2\sqrt{x(t - t_o)}}\right)\right], t > t_o \qquad (232\text{-}5.2)$$

Where $U(t-t_o)$ is the **unit-step function** defined as

$$
\begin{aligned}
U(t\text{-}t_o) \qquad &= 0 \qquad\qquad 0 < t < t_o \\
&= 1 \qquad\qquad t > t_o
\end{aligned}
$$

6.5. Exercise on using Laplace Transformation in solving Linear Partial Differential Equations

(1) Solve the partial differential equation

$$U_{xx}(x,t) + U_{tx}(x,t) - 2\ U_{tt}(x,t) = 0 \qquad\qquad (x > 0\ ,\ t > 0\ )$$

subject to the following boundary conditions

$U(x,0) = U_t(x,0) = 0$
$U(x,t) = 0$ as $x \to \infty$
$U(0,t) = F(t)$

(2) Solve the following boundary value problem

$$Y_x(x,t) + x\ Y_t(x,t) = 0$$

subject to the following boundary conditions

$Y(x,0) = 0$
$Y(0,t) = t$

(3) Solve the partial differential equation

$$Y_{tt}(x,t) = a^2\ Y_{xx}(x,t) \qquad\qquad x > 0,\ t > 0$$

subject to the following boundary conditions

$Y(x,0) = 0$
$Y_t(x,0) = -u_o$
$Y(0,t) = 0$
$Y_x\ (x,t) = 0$ as $x \to \infty$

(4) Find a bounded solution of

$$x\frac{\partial U}{\partial x} + \frac{\partial U}{\partial y} = xe^{-y} \qquad\qquad 0 < x < 1,\ y > 0$$

Which satisfies $U(x,0) = x$, $\qquad\qquad 0 < x < 1$

(5) Solve the equation

$$\frac{\partial U}{\partial t} + x\frac{\partial U}{\partial x} + U = x \qquad\qquad x > 0,\ t > 0$$

with the boundary conditions

$U(x,0) = 0$
$U(0,t) = 0$

(6) Show that *in* solving the partial differential equation

$$\frac{\partial^2 Y}{\partial t^2} = a^2 \frac{\partial^2 Y}{\partial x^2}$$

We can solve the equation

$$\frac{\partial^2 Y}{\partial t^2} = \frac{\partial^2 Y}{\partial x^2}$$

and then in the result we replace t by at.

(7) Solve the equation

$$\frac{\partial^2 Y}{\partial x^2} = 16 \frac{\partial^2 Y}{\partial t^2} \qquad\qquad x > 0 , t > 0$$

subject to the following boundary conditions

$Y(x,0) = 0$
$Y_t(x,0) = -1$
$Y(0,t) = t^2$
$Y_x (x,t)$ exists as $x \to \infty$ for fixed $t > 0$.

(8) Solve the equation

$$\frac{\partial Y}{\partial x} + 4 \frac{\partial Y}{\partial t} = -8t \qquad\qquad x > 0 , t > 0$$

subject to the following boundary conditions

$Y(x,0) = 0$
$Y(0,t) = 2 t^2$

(9) Solve the equation

$$\frac{\partial Y}{\partial x} + 2 \frac{\partial Y}{\partial t} = 4t \qquad\qquad x > 0 , t > 0$$

subject to the boundary conditions

$Y(x,0) = 0$
$Y(0,t) = 2 t^3$

(10)) Solve the equation

$$\frac{\partial^2 Y}{\partial t^2} = 4\frac{\partial^2 Y}{\partial x^2} \qquad\qquad x > 0, t > 0$$

subject to the boundary conditions

$Y(x,0) = 0$
$Y_t(x,0) = 2$
$Y(0,t) = \sin t$
$Y_x(x,t)$ exists as $x \to \infty$ for fixed $t > 0$.

Answers

(1) $U(x,t) = 0$ $0 < t < 2x$
          $U(x,t) = F(t - 2x)$ $t > 2x$]

(2) $Y(x,t)$ $= 0$ $0 < t < x^2/2$
                      $= t - x^2/2$ $t > x^2/2$]

(3) $Y(x,t)$ $= -u_0$ $0 < t < x/a$
                      $= -u_0 x/a$ $t > x/a$]

(4) $U(x,y) = xe^{-y}(1 + y)$

(5) $U(x,t) = U(x,t) = \frac{1}{2}x\left(1 - e^{-2t}\right)$

(6) Proof required

(7) $Y(x,t)$ $= -t$ $0 \le t \le 4x$
                      $= (t - 4x)^2 - 4x$ $t \ge 4x$

(8) $Y(x,t)$ $= -t^2$ $0 < t < 4x$
                      $= -t^2 + 3(t - 4x)^2$ $t > 4x$

(9) $Y(x,t)$ $= t^2$ $0 \le t \le 2x$
                      $= t^2 + 2(t - 2x)^3 - (t - 2x)^2$ $t \ge 2x$

(10) $Y(x,t)$ $= 2t$ $0 \le t \le x/2$
                      $= 2t + \sin(t - x/2) - 2(t - x/2)$ $t \ge x/2$

312

Index

www.ingramcontent.com/pod-product-compliance
Lightning Source LLC
Chambersburg PA
CBHW051210170526
45166CB00005B/1825